California Civil Seismic Principles

Practice Exams

Twelfth Edition

Majid Baradar, PE

PPI2PASS.COM
A **KAPLAN** COMPANY

Report Errors for This Book

PPI is grateful to every reader who notifies us of a possible error. Your feedback allows us to improve the quality and accuracy of our products. Report errata at **ppi2pass.com**.

CALIFORNIA CIVIL SEISMIC PRINCIPLES PRACTICE EXAMS
Twelfth Edition

Current release of this edition: 3

Release History

date	edition number	revision number	update
Mar 2018	12	1	New edition. Code updates. New content. Copyright update.
Jun 2019	12	2	Minor cover update.
Jul 2020	12	3	Code updates.

PPI
ppi2pass.com

ISBN: 978-1-59126-568-9

I dedicate this book to my son, Salar,
who is my pride and joy.

Table of Contents

Preface and Acknowledgments

I wrote *California Civil Seismic Principles Practice Exams* to help you prepare for the California Civil: Seismic Principles exam, which is administered by the California Board for Professional Engineers, Land Surveyors, and Geologists (BPELSG). This book will give you an opportunity to (1) evaluate the breadth and depth of your exam preparedness and (2) plan further study and review based on your strengths and weaknesses. The problems and solutions in this book present the basic concepts of seismic design fundamentals and demonstrate how relevant codes impose seismic considerations on the engineering design of structures.

To ensure that this twelfth edition would accurately reflect the topics on the exam, I reviewed all of the problems against the exam's test plan, which is provided on the BPELSG website. The test plan defines the fundamental principles, tasks, and elements of knowledge that are required to pass the exam. A link to the test plan is provided at **ppi2pass.com/CAspecial**.

I would like to thank Michael R. Lindeburg, PE, for contributing the "Definitions of Rigidity Used in This Book" section.

I also owe my most sincere gratitude to the PPI staff who worked on this new edition: Steve Buehler, director of acquisitions; Leata Holloway, senior acquisitions editor; Grace Wong, director of publishing services; Sam Webster, publishing systems manager; Nancy Peterson, editorial project manager; Gabby Howitz and Scott Rutherford, typesetters; Kim Wimpsett, proofreader; Tom Bergstrom, production associate and technical illustrator; Cathy Schrott, production services manager; Ellen Nordman, publishing systems specialist; and Jenny Lindeburg King, editor-in-chief. It has been my pleasure and great fortune to be able to collaborate with such an incredible publishing team.

I owe my loving thanks to my wife, Mitra Baradar, and my son, Salar Baradar. Without their encouragement and understanding, it would have been impossible for me to complete this book's first edition. My special gratitude is due to my mother, brothers, my sister, and their families for their loving support.

I value your suggestions and would appreciate your feedback regarding any errors or ambiguities in this book. You can submit comments using PPI's errata website at **ppi2pass.com/errata**. Finally, I offer this new edition to all engineers and professionals in the same spirit as I did earlier editions—to contribute to public safety in the event of an earthquake. I hope you find *California Civil Seismic Principles Practice Exams* valuable and rewarding—your career advancement is my sincerest wish. Good luck in your career!

Majid Baradar, PE

Introduction

ABOUT THIS BOOK

California Civil Seismic Principles Practice Exams contains two practice exams with problems that follow the multiple-choice format and the scope of topics on the California Civil: Seismic Principles exam. Like the actual exam, each practice exam contains 55 multiple-choice problems. The problems are similar in difficulty and complexity to actual exam problems. The practice exams are designed to be taken in the same length of time as the exam (i.e., two and a half hours). To simulate the flagging function on the computer-based test (CBT) exam, the answer sheets in this book also include a flag icon that you can use to identify the problems you want to come back to. Answer keys allow you to quickly score your exams, and complete solutions allow you to review and compare your own solving processes to identify errors and learn efficient solving approaches. Solutions and answer options are presented in customary U.S. units, with SI units provided where applicable.

This book uses the 2018 IBC and the 2016 ASCE/SEI7 when referencing code provisions in problems and solutions, unless a more stringent statutory requirement exists. (For example, Title 24 of the California Code of Regulations requires that schools and hospitals be operational after an earthquake, and this requirement supersedes lesser code requirements.) Since the 2018 IBC contains the same material and section numbers as the 2019 *California Building Code* (CBC), you may reference either the 2018 IBC or the 2019 CBC.

ABOUT THE CALIFORNIA CIVIL: SEISMIC PRINCIPLES EXAM

If you are applying to be a civil professional engineer in California and Guam, you must take and pass the California Civil: Seismic Principles exam to become a licensed civil engineer. The exam is offered year round on a quarterly basis. The exam is open-book and covers the content areas as outlined in the exam's test plan. You may only bring in as many reference materials as one box and one trip will permit. The exam is two and a half hours. For each problem, you will be asked to select the best answer from four options. Though you can answer problems at your own pace and in any order you choose, plan to spend an average of two and a half minutes on each problem. There is no penalty for guessing. Only correctly answered problems will be counted toward your score.

The California Civil: Seismic Principles exam is a CBT exam administered at a Prometric testing center. Navigation through the exam using the CBT interface is fairly standard. While you don't have to move sequentially through the exam, you can if you wish to. You can navigate to a specific problem. You can skip a problem or flag it for later review. You can return to any problem and change your answer. The onscreen index (listing, directory, etc.) can be used to see the problems that have been answered, flagged for review, or skipped. In some cases, you may have to toggle back and forth between a problem and an on-screen illustration if the illustration takes up too much space. A timer on the screen indicates how much time remains. For more information about Prometric and the exam site, review the Civil Seismic: Principles Exam candidate information bulletin (CIB) available on the Board for Professional Engineers, Land Surveyors, and Geologists (BPELSG) website.

HOW TO USE THIS BOOK

Prior to taking these practice exams, locate and organize relevant resources and materials as if you are taking the actual exam. You should also have the IBC and ASCE/SEI7 to be able to easily and quickly cross-reference the exam-adopted codes. Refer to the Codes and References Used in This Book section for guidance on other references you may wish to have during the exam.

The two practice exams in this book allow you to structure your exam preparation in a way that is best for you. For example, you might choose to take one exam as a pretest to assess your knowledge and determine the areas in which you need more review, and then take the second after you have completed additional study. Alternatively, you might choose to use one exam as a guide for how to solve different types of problems, reading each problem and solution in kind, and then use the second exam to evaluate what you learned.

However, these exams will be most useful if you treat them in this book as you would your actual exam. Do not read the problems ahead of time, and do not look at the solutions until you've answered all problems. Taking the practice exams in this book under the same time constraints and with the same reference material as the actual exam will help you assess your level of preparedness.

When you are ready to begin an exam, set a timer for two and a half hours. Use only the calculator and references you have gathered for use on the exam. For each exam, mark your answers on the appropriate answer sheet. Use the flag icon next to each problem on the answer sheet to mark those that you wish to return to.

Once you have completed a practice exam, check your answers using the answer keys. Make note of those areas you did well in and of those areas requiring more study. Review the solutions to any problems you've answered incorrectly or were unable to answer, and compare your problem-solving approaches against those given in the solutions.

The key to success on the actual exam is to practice solving as many problems as possible. This book will assist you with this objective.

HOW TO INTERPRET YOUR SCORE

The California Board for Professional Engineers, Land Surveyors, and Geologists has not divulged scoring methods or cut scores for the CBT exam; it instead reports results as either pass or fail on their website. Based on historical averages, the cut score was approximately 54%, with some problems weighted more heavily than others. For the practice exams in this book, use 70% as the target passing score.

If your score is at or below 70%, review the problems you had trouble with, and evaluate which areas require more study. If your score is greater than 70%, continue your review and concentrate on those areas where you may feel less confident.

Definitions of Rigidity Used in This Book

An understanding of structural rigidity is essential to applying seismic design principles. Most people are comfortable using the terms "rigid," "stiff," and "inflexible" as synonyms. However, engineers actually need to be able to quantify rigidity, since some structures are more rigid than others. Rigidity has traditionally been defined as "the reciprocal of deflection," although rigidity is seldom used in that context. Unfortunately for someone learning seismic design principles for the first time, the term "rigidity" has at least four different incompatible usages; and the codes and literature, which are written for practitioners, not students, do not distinguish between the usages and are often inconsistent. Without a clear understanding of the usages, the rigidity variable, R, is easily misused.

Absolute rigidity, R_{abs}, is 100% synonymous with stiffness. Hooke's law states that the deflection of a structure is proportional to the applied force, and the constant of proportionality is the *stiffness*, k. For example, the relationship between stiffness and the deflection, x, of a spring acted upon by a force, F, is

$$F = kx \quad \text{[Hooke's law]}$$

For seismic design use, Hooke's law would be rewritten as $V = R_{abs}\delta$ or $V = R_{abs}\Delta$, where V is the applied shear force and either δ or Δ is used to represent the deflection. Notice that R_{abs} has units of force/length (e.g., kips/in). Either the context or the units can be used to identify when the word "rigidity" should be taken to mean "stiffness." The usefulness of this definition of rigidity is that the deflection can be determined for any disturbing force (i.e., for any earthquake). The absolute rigidity of a member can be used to find the deflection from any force.

Hooke's law shows that rigidity has something to do with deflection, but the "reciprocal of deflection" definition does not follow from the mathematics. In order to use this definition, the force would have to be disregarded. For analysis (not for design) work, this definition could be used to compare the behavior (response) of different designs exposed to the same disturbing force. Basically, the force can be disregarded if it is a constant, and even then, the reciprocal of deflection can only be used to rate, rank, or describe the observed behavior of a specific structure under those specific conditions. Under those conditions, the *observed rigidity* could be defined as

$$R_{obs} = \frac{1}{\delta}$$

Observed rigidity can be distinguished from any other definition of rigidity by its units, which are 1/length (e.g., 1/in). Unfortunately, observed rigidity is often reported without units, which requires the practitioner to assume units or decide the units based on logic. Observed rigidity isn't particularly useful, as it can't be used to determine the applied force or the stiffness of a structure or the structure's response to a different disturbing force. Observed stiffness can be determined and reported (with units), but by itself, the observed stiffness of a single wall or structure can't be used for anything.

The *tabulated rigidity*, R_{tab}, combines the concepts of stiffness and observed rigidity. Tabulated rigidity is the reciprocal of deflection as calculated from mechanics of materials principles for an imaginary structure with arbitrary properties. For manual solutions, tabulated rigidity, as the name implies, is looked up in a table. Values can be calculated instead, but the calculations must consider the contributions of both moment and shear to deflection, and the equations are somewhat more complex than the simple elastic beam equations. For the convenience of anyone solving a seismic design problem manually, tabulated rigidity is most easily read from a table.

Not unexpectedly, the deflection depends on the applied force, the structure's material, and its dimensions, so the deflection equation contains terms for modulus of elasticity, E, moment of inertia, I, and length. The actual values of these terms are unknown, so values of tabulated rigidity represent the reciprocals of deflection for an arbitrary structure based on a convenient but arbitrary force, dimensions, and material properties. For example, a thickness of 1 in, a force of 100,000 lbf, and a modulus of elasticity of 1,000,000 psi are used to calculate the tabulated rigidities in most tabulations. Other values could be used, but these are the most common. Since convenient but arbitrary values are used, the deflection (and its reciprocal, the tabulated rigidity) clearly cannot be applied to any specific structure.

Since tabulated rigidity has the same units as observed rigidity, 1/length, tabulated rigidity suffers from the same limitation as observed rigidity (i.e., it is largely descriptive). However, by itself, it is even less useful than the $1/\delta$ calculation for observed rigidity (which was based on an observed deflection of an actual structure); it is based on arbitrary values of an imaginary structure. Individual values of tabulated rigidity can't be applied to any specific structure. It would be impossible, for example, to obtain a value of R_{tab} for use in determining the deflection of a shear wall.

Tabulated rigidity's usefulness comes from an understanding of how seismic design is practiced. Seismic designers essentially know the force that the building must withstand. They know the expected ground acceleration, and as the design progresses, they have increasingly good estimates of the building's final mass. $F = ma$ and its code-related equivalent ($V = CW$) tell them what total force the building will be exposed to. This force is resisted by individual pieces of the structure. Seismic designers don't design buildings; they design the pieces of buildings—the walls, columns, roofs, floors, and connections. Seismic designers need to know how much of the applied force each piece resists, and they need a way to distribute the total earthquake force among the various pieces. They do this by determining the relative strength of each piece. The relative strength of a piece is the strength relative to all the force-resisting members. The term "relative" implies more than one force-resisting member.

The relative strength of a particular component (e.g., one of several shear walls) is referred to as the *relative rigidity*, R_{rel}. A single member may have its own relative rigidity, but the relative rigidity can't be determined without knowledge of all the force-resisting members. The relative rigidity of a particular component, i, among n components that share the earthquake force, V, is calculated from a ratio of the tabulated rigidities.

$$R_{\text{rel},i} = \frac{R_{\text{tab},i}}{\displaystyle\sum_{i=1}^{n} R_{\text{tab},i}}$$

Even though the individual values of tabulated rigidity are based on arbitrary values of force, material properties, and dimensions, these arbitrary values cancel out since they are represented in both the numerator and denominator values. Only the relevant dimensions (base and height) used to look up the R_{tab} values implicitly remain. The force resisted by component i is

$$V_i = R_{\text{rel},i} V = \frac{R_{\text{tab},i}}{\displaystyle\sum_{i=1}^{n} R_{\text{tab},i}} V$$

Relative rigidity is unitless. It is a fraction less than 1.0—the fraction of the total earthquake force resisted by component i. The sum of all relative rigidities within a system of force-resisting members is 1.0.

$$\sum_{i=1}^{n} R_{\text{rel},i} = 1.0$$

Even though relative rigidity shares the common name "rigidity," it is inconsistent with all three other definitions of rigidity. It can't directly be used to calculate the deflection of an individual wall.

Unfortunately, unlike what is presented in this book, common usage does not distinguish between absolute rigidity, observed rigidity, and tabulated rigidity. In truth, the terms "absolute rigidity," "observed rigidity," and "tabulated rigidity" were made up specifically for this book; they don't exist in practice. And, although the term "relative rigidity" is used in practice, the term is sometimes applied incorrectly to the tabulated values. The correct definition and usage of the term "rigidity" must usually be determined from the context, units, and value. (A relative rigidity could never have a value greater than 1.0.) This determination is essential. None of the four types of rigidity can be substituted for the others in usage.

Codes and References Used in This Book

The information that was used to write and update this book was based on the exam's test plan at the time of publication. However, as with engineering practice itself, the California Civil: Seismic Principles exam is not always based on the most current codes or cutting-edge technology. Similarly, codes, standards, and regulations adopted by state and local agencies often lag issuance by several years. It is likely that the codes that are most current, the codes that you use in practice, and the codes that are the basis of your exam will all be different.

PPI lists on its website the dates and editions of the codes, standards, and regulations that the exam is based on. It is your responsibility to find out which codes are relevant to your exam. In the meantime, here are the codes and references that have been used in writing this book.

CODES AND REFERENCES

Alquist-Priolo Earthquake Fault Zoning Act *California Public Resources Code* 2013, Title 14, Chap. 7.5, §2621–2630

ATC-20 Procedures for Postearthquake Safety Evaluation of Buildings and *ATC-20-1 Field Manual: Postearthquake Safety Evaluation of Buildings* 1995

Board Rules and Regulations Relating to the Practices of Professional Engineering and Professional Land Surveying. *California Code of Regulations* 2014, Title 16, Div. 5, §400–476

Building Code Requirements and Specifications for Masonry Structures (TMS 402-16, TMS 602-16)

California Building Code (CBC) 2019, California Building Standards Commission

California Health and Safety Code 2014, Title 8

International Building Code (IBC) 2018, International Code Council

Field Act *California Building Code* 2013, Part I, Title 24, §4-301, *et seq.*

Minimum Design Loads for Buildings and Other Structures (ASCE/SEI7) 2016, American Society of Civil Engineers

NEHRP Recommended Seismic Provisions for New Buildings and Other Structures (FEMA P-1050) 2015, Building Seismic Safety Council

Professional Engineers' Act *California Business and Professions Code* 2014, Chap. 7, §6700–6799

Professional Land Surveyors' Act *California Business and Professions Code* 2014, Chap. 15, §8700–8805

Seismic Provisions for Structural Steel Buildings 2016, American Institute of Steel Construction (AISC 341)

Special Design Provisions for Wind and Seismic (AWC SDPWS) 2015

Practice Exam 1 Instructions

In accordance with the rules established by the California Board for Professional Engineers, Land Surveyors, and Geologists, you may use textbooks, handbooks, bound reference materials, and any approved battery- or solar-powered, silent calculator to work this examination. However, no blank papers, writing tablets, unbound scratch paper, or loose notes are permitted. Sufficient room for scratch work is provided in the Examination Booklet. You are not permitted to share or exchange materials with other examinees.

You will have $2\frac{1}{2}$ hours in which to work this examination. All questions must be worked correctly in order to receive full credit on the exam. There are no optional questions. Partial credit is not available. No credit will be given for methodology, assumptions, or work written in your Examination Booklet.

Record all of your answers on the Answer Sheet. No credit will be given for answers marked in the Examination Booklet. To simulate the flagging function on the computer-based test (CBT) exam, the answer sheet includes a flag icon that you can use to identify the problems you want to come back to.

If you finish early, check your work and make sure that you have followed all instructions. After checking your answers, you may turn in your Examination Booklet and Answer Sheet and leave the examination room. Once you leave, you will not be permitted to return to work or change your answers.

WAIT FOR PERMISSION TO BEGIN

Name: _____
       Last   First   Middle Initial

Examinee number: _____

Examination Booklet number: _____

California Civil: Seismic Principles Exam

Practice Exam 1

Practice Exam 1 Answer Sheet

1. Ⓐ Ⓑ Ⓒ Ⓓ ⚑	15. Ⓐ Ⓑ Ⓒ Ⓓ ⚑	29. Ⓐ Ⓑ Ⓒ Ⓓ ⚑	43. Ⓐ Ⓑ Ⓒ Ⓓ ⚑
2. Ⓐ Ⓑ Ⓒ Ⓓ ⚑	16. Ⓐ Ⓑ Ⓒ Ⓓ ⚑	30. Ⓐ Ⓑ Ⓒ Ⓓ ⚑	44. Ⓐ Ⓑ Ⓒ Ⓓ ⚑
3. Ⓐ Ⓑ Ⓒ Ⓓ ⚑	17. Ⓐ Ⓑ Ⓒ Ⓓ ⚑	31. Ⓐ Ⓑ Ⓒ Ⓓ ⚑	45. Ⓐ Ⓑ Ⓒ Ⓓ ⚑
4. Ⓐ Ⓑ Ⓒ Ⓓ ⚑	18. Ⓐ Ⓑ Ⓒ Ⓓ ⚑	32. Ⓐ Ⓑ Ⓒ Ⓓ ⚑	46. Ⓐ Ⓑ Ⓒ Ⓓ ⚑
5. Ⓐ Ⓑ Ⓒ Ⓓ ⚑	19. Ⓐ Ⓑ Ⓒ Ⓓ ⚑	33. Ⓐ Ⓑ Ⓒ Ⓓ ⚑	47. Ⓐ Ⓑ Ⓒ Ⓓ ⚑
6. Ⓐ Ⓑ Ⓒ Ⓓ ⚑	20. Ⓐ Ⓑ Ⓒ Ⓓ ⚑	34. Ⓐ Ⓑ Ⓒ Ⓓ ⚑	48. Ⓐ Ⓑ Ⓒ Ⓓ ⚑
7. Ⓐ Ⓑ Ⓒ Ⓓ ⚑	21. Ⓐ Ⓑ Ⓒ Ⓓ ⚑	35. Ⓐ Ⓑ Ⓒ Ⓓ ⚑	49. Ⓐ Ⓑ Ⓒ Ⓓ ⚑
8. Ⓐ Ⓑ Ⓒ Ⓓ ⚑	22. Ⓐ Ⓑ Ⓒ Ⓓ ⚑	36. Ⓐ Ⓑ Ⓒ Ⓓ ⚑	50. Ⓐ Ⓑ Ⓒ Ⓓ ⚑
9. Ⓐ Ⓑ Ⓒ Ⓓ ⚑	23. Ⓐ Ⓑ Ⓒ Ⓓ ⚑	37. Ⓐ Ⓑ Ⓒ Ⓓ ⚑	51. Ⓐ Ⓑ Ⓒ Ⓓ ⚑
10. Ⓐ Ⓑ Ⓒ Ⓓ ⚑	24. Ⓐ Ⓑ Ⓒ Ⓓ ⚑	38. Ⓐ Ⓑ Ⓒ Ⓓ ⚑	52. Ⓐ Ⓑ Ⓒ Ⓓ ⚑
11. Ⓐ Ⓑ Ⓒ Ⓓ ⚑	25. Ⓐ Ⓑ Ⓒ Ⓓ ⚑	39. Ⓐ Ⓑ Ⓒ Ⓓ ⚑	53. Ⓐ Ⓑ Ⓒ Ⓓ ⚑
12. Ⓐ Ⓑ Ⓒ Ⓓ ⚑	26. Ⓐ Ⓑ Ⓒ Ⓓ ⚑	40. Ⓐ Ⓑ Ⓒ Ⓓ ⚑	54. Ⓐ Ⓑ Ⓒ Ⓓ ⚑
13. Ⓐ Ⓑ Ⓒ Ⓓ ⚑	27. Ⓐ Ⓑ Ⓒ Ⓓ ⚑	41. Ⓐ Ⓑ Ⓒ Ⓓ ⚑	55. Ⓐ Ⓑ Ⓒ Ⓓ ⚑
14. Ⓐ Ⓑ Ⓒ Ⓓ ⚑	28. Ⓐ Ⓑ Ⓒ Ⓓ ⚑	42. Ⓐ Ⓑ Ⓒ Ⓓ ⚑	

Practice Exam 1

$$y = y_1 + \frac{x - x_1}{x^2 - x_1}(y_2 - y_1)$$

1. Which California law requires the State Geologist to establish regulatory zones around all known active faults?

(A) Seismic Zone Act

(B) Alquist-Priolo Act

(C) Riley Act

(D) Special Studies Zones Act

2. The design earthquake motion per ASCE/SEI7 provisions corresponds to

(A) 2/3 times the maximum considered earthquake design motion

(B) 2/3 times the earthquake that has a 10% probability of being exceeded in 50 years

(C) 3/2 times the maximum considered earthquake ground motion

(D) 3/2 times the earthquake that has a 10% probability of being exceeded in 50 years

3. The building shown has walls of uniform thickness.

W = 120 kips (530 kN)
W = 130 kips (580 kN)
W = 160 kips (710 kN)
W = 100 kips (440 kN)

What type of vertical structural irregularity is probable?

I. soft story

II. mass (weight) irregularity

III. in-plane discontinuity

(A) I only

(B) II only

(C) I and II only

(D) I, II, and III

4. A 155 ft (47 m) steel building in Los Angeles will be braced laterally by a steel eccentrically braced frame structure. The structure will be utilized as a communication center that can respond in emergencies. The geotechnical engineer estimates the S_1 and S_S values are 0.2 and 0.5, respectively. The design base shear using the equivalent lateral-force procedure is most nearly

(A) $0.42\,W$

(B) $0.044\,W$

(C) $0.057\,W$

(D) $0.088\,W$

5. The design base shear for an 80 ft (24 m) structure with eight stories of equal height and floor weight is 49,500 lbf (220 000 N). The fundamental period for this structure is 0.68 sec. The distributed base shear at level three is most nearly

(A) 4000 lbf (18 000 N)

(B) 4100 lbf (19 000 N)

(C) 7800 lbf (34 000 N)

(D) 8500 lbf (38 000 N)

6. Which geotechnical conditions can damage even well-designed structures?

    I.   large settlement or lateral spreading of soil beneath structures

    II.   ground movements associated with surface rupture

    III.   slope failure

(A)   I only

(B)   II only

(C)   I and II only

(D)   I, II, and III

7. Which ASCE/SEI7 formula should be used to determine the total design lateral seismic force on elements of nonstructural components and equipment supported by structures?

(A) $V = \dfrac{S_{DS}}{\dfrac{R}{I_e}} W$

(B) $F_p = \left(\dfrac{0.4 a_p S_{DS} W_p}{\dfrac{R_p}{I_p}} \right)\left(1 + 2\dfrac{z}{h} \right)$

(C) $V = \dfrac{w_x h_x^k}{\displaystyle\sum_{i=1}^{n} w_i h_i} W$

(D) $V = \dfrac{F S_{DS}}{R} W$

8. In California, which types of building construction have generally had a good seismic performance record?

    I.   unreinforced masonry (URM) bearing wall buildings

    II.   steel and concrete frames with URM infill walls

    III.   nonductile concrete frames

(A)   I only

(B)   III only

(C)   I and II only

(D)   none of the above

9. Three special steel moment-resisting frame structures are shown. For all structures, the design spectral response acceleration parameters are $S_{DS} = 0.75$, and $S_{D1} = 0.6$, the seismic design category is F, and the risk category is III. Based on ASCE/SEI7 requirements, which structures can be designed using the equivalent lateral-force procedure for its seismic force-resisting systems?

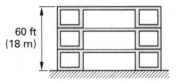

60 ft
(18 m)

horizontal irregular structure, type 4

structure I

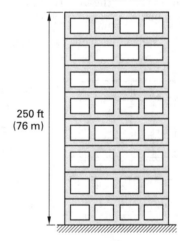

160 ft
(49 m)

regular structure

structure II

250 ft
(76 m)

regular structure

structure III

(A)   none

(B)   I only

(C)   I and II only

(D)   I, II, and III

10. In a building, an adequate load path is essential for seismic resistance. In a lateral-force resisting system, the complete path of the lateral force should be in what order?

I.   The vertical elements transfer the lateral forces into the foundation, and the foundation transfers them to the ground.

II. The diaphragms distribute lateral forces to vertical elements of the building.

III. Lateral forces are transferred through structural connections to the horizontal diaphragm.

(A)   through III only

(B)   from III to II

(C)   from II to III to I

(D)   from III to II to I

11. Deficient shear capacity of an existing wood structural panel diaphragm can be improved and strengthened by which techniques?

I.   providing additional nailing

II. adding a new layer of wood structural panels

III. reducing the diaphragm's span through the addition of new shear walls or braced frames

(A)   III only

(B)   I and III only

(C)   I, II, and III

(D)   none of the above

12. The following documents contain factual data only. Which documents should NOT bear the seal and signature of a registered civil engineer?

I.   project background and information sheets

II. reports on gathered environmental data

III. draft/incomplete documents

(A)   III only

(B)   I and II only

(C)   I and III only

(D)   I, II, and III

13. For a one-story wood-frame building with a wood structural panel roof diaphragm, consider two walls: the west wall at midspan and the south wall at the intersection of lines X and 1. The redundancy/reliability factor, ρ, is 1.0.

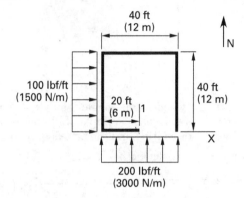

Which statement is correct?

(A)   The chord forces in both walls are the same.

(B)   The chord force in the south wall is twice the chord force in the west wall.

(C)   The chord force in the south wall is half the chord force in the west wall.

(D)   The chord force in the south wall is zero.

14. The largest earthquake ground motion that is expected to occur sometime in the life of a structure to be built at a specific site is the maximum

(A)   probable earthquake

(B)   considered earthquake

(C)   predictable earthquake

(D)   theoretical earthquake

15. The natural frequency, f, of system I is 0.5 Hz, and the natural period, T, of system II is 2 sec. Systems I and II are single-degree-of-freedom systems. Which statement is correct?

(A) $T_I > T_{II}$ and $f_I > f_{II}$

(B) $T_I = T_{II}$ and $f_I = f_{II}$

(C) $T_I > T_{II}$ and $f_I < f_{II}$

(D) $T_I < T_{II}$ and $f_I < f_{II}$

16. A 120 ft (36 m), eight-story structure consists completely of steel moment-resisting frames. The structure has an S_{D1} value of 0.2. Most nearly, what is its maximum fundamental period?

(A)   1.0 sec (1.0 s)

(B)   1.3 sec (1.3 s)

(C)   1.5 sec (1.5 s)

(D)   1.9 sec (1.9 s)

$T = C_u T_a \Rightarrow maximum$

17. An existing wood-frame, single-family home in Los Angeles has a crawl space (cripple wall) between the raised concrete foundation and the first floor as shown.

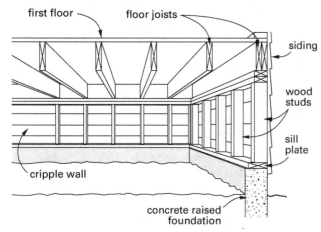

What strengthening measures will enhance the seismic resistance of this deficient cripple wall?

I.   using anchor bolts to connect the sill plates and the foundation

II.   using hold-downs to anchor the wood-stud walls to the foundation

III.   nailing wood structural panels on the inside of the cripple studs

(A)   I only

(B)   II only

(C)   I and II only

(D)   I, II, and III

18. Two site class C structures with no structural irregularities are shown. The mapped acceleration parameters are $S_S = 0.75$ and $S_1 = 0.40$. The total weight of both buildings is equal.

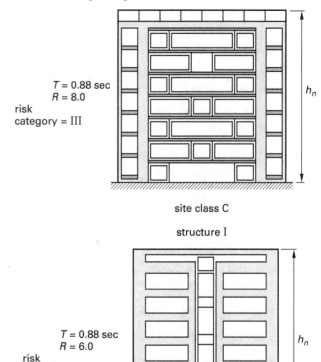

site class C

structure I

site class C

structure II

Which statement best describes the value of the base shear, V, for the structures?

(A)   Structure I has a smaller base shear than structure II.

(B)   Structure II has a smaller base shear than structure I.

(C)   Structure I and structure II have equal base shears.

(D)   There is not enough information.

check all
eqs!

19. Which statement(s) is/are correct?

   I. Diaphragms are horizontal subsystems.

   II. Diaphragms transmit lateral forces to the vertical resisting elements.

   III. Diaphragms typically consist of the floors and roofs of a building.

(A) II only

(B) I and II only

(C) II and III only

(D) I, II, and III

Diaphragms are systems!

20. Which statement is true for both rigid diaphragms and flexible diaphragms?

(A) They distribute lateral forces in proportion to the rigidities of vertical resisting elements.

(B) They transmit torsion to the vertical resisting elements.

(C) They distribute lateral story shears to vertical resisting elements.

(D) They distribute lateral forces in proportion to the tributary area of vertical resisting elements.

21. As defined in the Professional Engineers Act, the phrase "responsible charge of work" directly relates to a professional engineer when he or she

   I. makes engineering decisions or recommendations

   II. judges the validity of professional engineering work

   III. accepts financial liability

(A) I only

(B) III only

(C) I and II only

(D) I, II, and III

22. Three large structures with different types of structural systems are shown. Which structure lacks a complete vertical load-carrying space frame?

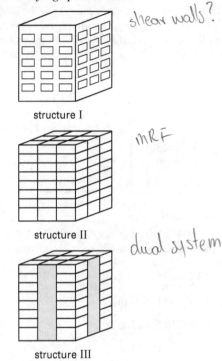

shear walls?

structure I

MRF

structure II

dual system

structure III

(A) I only

(B) II only

(C) III only

(D) none of the above

23. The Field Act was passed following severe damage to many structures during the Long Beach earthquake of 1933. This act requires special seismic design for which structures?

(A) public school buildings

(B) private school buildings

(C) hospitals

(D) both option A and option B

24. The wood structural panel roof diaphragm of a residential one-story building is shown in plan view. The lateral force is in the east-west direction. The calculated roof diaphragm shear capacity is 300 lbf/ft (4400 N/m). The roof diaphragm is adequately anchored to the shear wall, and the reliability/redundancy factor, ρ, is 1.0.

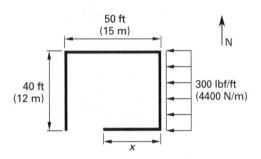

To have a drag strut force of 2000 lbf (8900 N), the south shear wall panel should most nearly be

(A)   20 ft (8.0 m)

(B)   30 ft (10 m)

(C)   40 ft (12 m)

(D)   50 ft (14 m)

25. The lifter system shown is capable of moving loads up and down. The response spectra and a table illustrating the variation of the system's natural period, T, versus mass height, h, are given.

W	h	T
	100 ft (30 m)	2.0 sec
20 kips	75 ft (23 m)	1.2 sec
(9000 kg)	50 ft (15 m)	0.8 sec
	25 ft (7.6 m)	0.3 sec

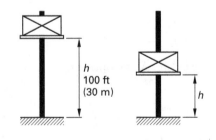

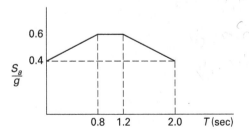

The maximum overturning moment at the base is most nearly

(A)   500 ft-kips (600 kN·m)

(B)   600 ft-kips (700 kN·m)

(C)   900 ft-kips (1200 kN·m)

(D)   2000 ft-kips (2400 kN·m)

26. In high seismic areas on site classes A and B, what is the most effective technique to reduce the acceleration, and therefore, the base shear imposed on the structural system of a building?

(A)   Installing seismic base isolators at the base of the building.

(B)   Increasing the height of the building.

(C)   Modifying the existing structural system.

(D)   For bedrock soil profile types S_A and S_B, acceleration should not be decreased.

27. A wood-frame shear wall is shown in elevation. The shear wall is part of a one-story structure with a flexible roof diaphragm. The shear wall panels have the same thickness, and each panel has a shear capacity of 500 lbf/ft (7300 N/m).

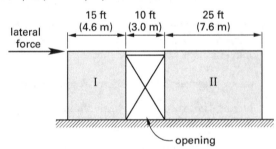

The maximum applied lateral force to the wall is most nearly

(A)   15,000 lbf (67 000 N)

(B)   20,000 lbf (89 000 N)

(C)   25,000 lbf (110 000 N)

(D)   30,000 lbf (130 000 N)

28. A registered professional civil engineer is competent and proficient in highway and transportation design. Which structures can this professional engineer design?

   I. a high-rise structure

   II. a hospital under the direction of a registered structural engineer

   III. a highway bridge structure

(A) I only

(B) II only

(C) III only

(D) I, II, and III

29. In the event of an earthquake, the primary goal of seismic design is to

   I. prevent loss of life

   II. prevent major structural failures

   III. preserve property

(A) I only

(B) I and II only

(C) I and III only

(D) I, II, and III

30. The R factor accounts for the inherent overstrength and global ductility (energy absorption capacity) of a structure. The value of R is based on

   I. selection of structural systems

   II. combinations of structural systems along the same axes

   III. vertical combinations of structural systems

(A) I only

(B) I and II only

(C) II and III only

(D) I, II, and III

31. Which equation would be used to calculate the minimum base shear for a building where $S_1 = 0.75$?

(A) $V = \left(\dfrac{S_{DS}}{\dfrac{R}{I_e}} \right) W$

(B) $V = \left(\dfrac{0.5 S_1}{\dfrac{R}{I_e}} \right) W$

(C) $V = \left(\dfrac{S_{D1}}{T \left(\dfrac{R}{I_e} \right)} \right) W$

(D) $V = \left(\dfrac{S_{D1} T_L}{T^2 \left(\dfrac{R}{I_e} \right)} \right) W$

32. Based on ASCE/SEI7 requirements, which illustrated cases have correctly accounted for the accidental eccentricity in the direction of the applied loading?

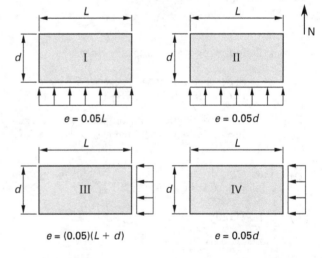

(A) I only

(B) III only

(C) I and IV only

(D) II and III only

33. Structures are most likely to be damaged by liquefaction if they are supported on which foundation and soil conditions?

(A) isolated spread footings supported on a soil profile consisting of predominantly saturated cohesionless soil

(B) piles or drilled piers that extend through a soil profile consisting of deep saturated cohesionless soil and are supported on rock-like materials

(C) continuous spread footings supported on a soil profile containing a deep clay layer

(D) none of the above

34. A multi-story structure uses rigid diaphragms to transfer loads to the three different building systems illustrated.

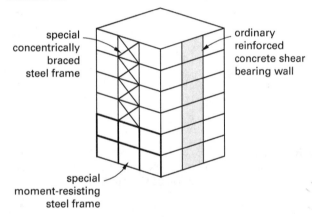

Which of the following statements is correct?

I. The structure would be designated as a triple system according to ASCE/SEI7.

II. The structure would NOT be permitted in seismic design category D according to ASCE/SEI7.

III. The structure needs to be checked for accidental torsion according to ASCE/SEI7.

(A) I and II only

(B) I and III only

(C) II and III only

(D) I, II, and III

35. The bridge structure shown is subjected to a lateral load. Columns are square, the modulus of elasticity, E, is 29×10^6 lbf/in^2 (2.0×10^5 MPa), and the moment of inertia, I, per column is 19.5 in^4 (8.1×10^{-6} m^4). The supporting columns A and C are fixed at the tops and

bottoms. The center column, B, is fixed at the top and pinned at the bottom.

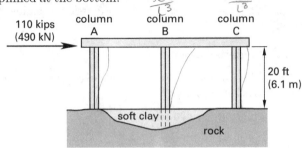

The resisting force in column B will be most nearly

(A) 10 kips (50 kN)

(B) 20 kips (100 kN)

(C) 40 kips (160 kN)

(D) 50 kips (200 kN)

36. Which registered professionals can design and sign plans for a two-story single-family dwelling and a two-story four-unit condominium complex of wood-frame construction?

I. structural engineers

II. civil engineers

III. architects

(A) I only

(B) I and II only

(C) I and III only

(D) I, II, and III

37. What are the values of the site coefficients, F_a and F_v, for a building in site class D, and where $S_S = 0.9$ and $S_1 = 0.5$?

(A) 1.0 and 1.0

(B) 1.0 and 1.14

(C) 1.14 and 1.0

(D) 1.14 and 1.8

38. The stability coefficient is calculated using the equation shown.

$$\theta = \frac{P_x \Delta}{V_x h_{sx} C_d}$$

The *P*-delta effects on story shears and moments need NOT be considered when θ is

(A) less than or equal to 0.10

(B) less than $\theta_{\max} = \dfrac{0.5}{\beta C_d} \le 0.25$

(C) greater than $\theta_{\max} = \dfrac{0.5}{\beta C_d} \le 0.25$

(D) less than 1.0

39. A one-story masonry shear wall structure with a rigid diaphragm is shown in plan view. The table shown gives the tabulated rigidity of each shear wall, R_{tab}; the center of mass, CM; the center of rigidity, CR; and the distance from the center of rigidity perpendicular to the wall, d. The base shear is 50,000 lbf (220 000 N) for the north-south direction. The shown locations of CM and CR are appropriate.

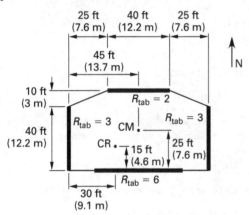

wall	$R_{\text{tab},x}$	$R_{\text{tab},y}$	d_x	d_y	$R_{\text{tab}}d$	$R_{\text{tab}}d^2$
north	2	–	–	35 ft (10.6 m)	70 ft (21.2 m)	2450 ft² (225 m²)
south	6	–	–	−15 ft (4.6 m)	−90 ft (27.6 m)	1350 ft² (127 m²)
east	–	3	60 ft (18.3 m)	–	180 ft (54.9 m)	10,800 ft² (1000 m²)
west	–	3	−30 ft (9.1 m)	–	−90 ft (27.3 m)	2700 ft² (248 m²)

For the east wall, the total force due to shear and torsion is most nearly

(A) 10,000 lbf (44 500 N)

(B) 18,000 lbf (80 000 N)

(C) 25,000 lbf (111 000 N)

(D) 35,000 lbf (155 000 N)

40. To use the title "Professional Engineer" in the professional practice of rendering civil engineering services, an engineer must

  I. be registered as a professional engineer

 II. be competent and qualified according to the Board's rules and regulations

III. have a degree in civil engineering

(A) I only

(B) II only

(C) I and II only

(D) I, II, and III

41. Elements connecting smaller portions of a structure to the remainder of the structure must have a design strength of at least

(A) $0.2 S_{DS}$ times the weight of the smaller portion

(B) 5% of the smaller portion weight

(C) $0.3 S_{DS}$ times the weight of the smaller portion

(D) 10% of the smaller portion weight

42. Two lateral loads are applied to the small moment frame as shown. A distributed load of 2 kips/ft (29 kN/m) is applied across the length of the horizontal members. Columns A and B are pinned at both sides.

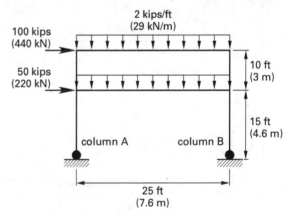

The axial force in column A due to the applied loads is most nearly

(A) 50 kips (220 kN)

(B) 80 kips (350 kN)

(C) 130 kips (580 kN)

(D) 180 kips (800 kN)

43. A connection between a steel deck diaphragm and a concrete or masonry wall is shown. The decking is filled with reinforced concrete. The steel deck is supported by a steel ledger and is bolted to the wall.

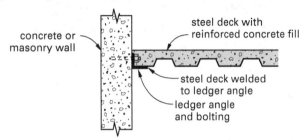

Which illustrated technique best increases shear capacity and anchorage of the connection to the wall?

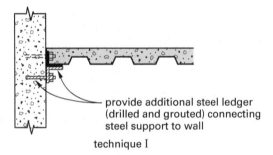

technique I

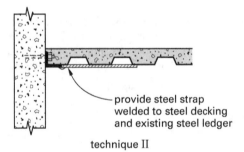

technique II

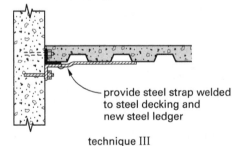

technique III

(A) I

(B) II

(C) III

(D) This connection has no deficiency.

44. An engineering firm is designing a residential building in a high seismic area. Various building materials have been studied for its construction. Considering material costs and the speed of construction, which selection is the most appropriate for the least expensive and earliest completion of the project?

(A) concrete

(B) masonry

(C) steel

(D) wood

45. Which earthquake engineering tasks can be performed by structural computer programs?

   I.   static/dynamic/P-delta analysis

   II.   time history analysis

   III.   response spectrum analysis

(A) I only

(B) I and II only

(C) I and III only

(D) I, II, and III

46. The maximum diaphragm chord force and strut force similarly develop at what location?

   I.   the location of maximum moment

   II.   the ends of the boundary members

   III.   the points of discontinuity in the plan

(A) I only

(B) III only

(C) I, II, and III

(D) none of the above

47. On the Richter scale, a magnitude increase of one unit (i.e., one whole number) represents how much of an increase in radiated energy?

(A) equal energy

(B) 10 times more energy

(C) 32 times more energy

(D) 100 times more energy

48. What is the response modification factor of the structure shown for an earthquake in the *y*-direction?

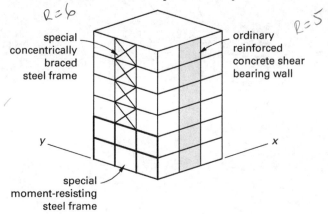

R = 6

R = 5

special concentrically braced steel frame

ordinary reinforced concrete shear bearing wall

y

x

special moment-resisting steel frame

(A)   4

(B)   6

(C)   7

(D)   8

49. In accordance with the appropriate IBC table, for a one-story, wood-frame residential building with the given characteristics, what should be the maximum allowable shear for seismic forces for the building's wood structural panel shear walls?

3/8 in (10 mm) thick structural I panels

1½ 16 gage

wood structural panels applied directly to the framing members

blocking required

staple spacing at panel edges is 3 in (76 mm)

12 in (305 mm) on center (o.c.) field stapling

studs spaced at 24 in (610 mm) o.c.

(A)   155 lbf/ft (2260 N/m)

(B)   315 lbf/ft (4600 N/m)

(C)   375 lbf/ft (5470 N/m)

(D)   475 lbf/ft (6930 N/m)

50. According to the IBC, which of the following statements is true?

a

   I.   An approved agency is an institution approved by the local building official or authority having jurisdiction that performs materials testing or inspection of civil engineering projects.

per material or product

  II. An inspection certificate is a document signed by a qualified inspector at the completion of a construction project to signify that the building aligns with the plans and specifications.

 III. Continuous special inspection must be performed by a part-time inspector who is present when and where the work to be inspected is being done.

 IV. Structural observation by a registered design professional may be substituted for the work done by the inspector.

NOT APP.

(A)   I only

(B)   II only

(C)   I and III only

(D)   II and IV only

51. According to the IBC, which of the following construction types requires special inspection?

   I.   vertical masonry foundation elements

  II. anchors cast in concrete

 III. cold-formed steel deck

(A)   I only

(B)   II only

(C)   III only

(D)   I, II, and III

52. A diaphragm structure is anchored to a concrete shear wall. Which of the following must be known to design the diaphragm anchors?

   I.   out-of-plane loads

  II. weight of the tributary portion of the wall

 III. in-plane shear loads

 IV. redundancy factor

(A)   I and II only

(B)   I and IV only

(C)   II and III only

(D)   I, II, and III

53. A four-story building has the lateral strength values shown.

200 kips (890 kN)

220 kips (980 kN)

250 kips (1100 kN)

180 kips (800 kN)

Which type of structural irregularity does the building most likely have?

(A) stiffness-soft story irregularity

(B) stiffness-extreme soft story irregularity

(C) weak story irregularity

(D) extreme weak story irregularity

54. According to ASCE/SEI7, which of the following is NOT used to determine the base shear of a structural system?

(A) response modification coefficient

(B) overstrength factor

(C) deflection amplification factor

(D) redundancy factor

55. A 160 ft (50 m), 14-story building has an eccentrically braced steel frame and a one-second design spectral response acceleration parameter, S_{D1}, of 0.5. According to computer modeling and analysis, the building's fundamental period of vibration is 2.2 sec. What is most nearly the fundamental period that should be used to determine the building's seismic response coefficient?

(A) 1.0 sec

(B) 1.4 sec

(C) 1.9 sec

(D) 2.2 sec

STOP!

DO NOT CONTINUE!

This concludes the California Civil: Seismic Principles Exam. If you finish early, check your work and make sure that you have followed all instructions. After checking your answers, you may turn in your examination booklet and answer sheet and leave the examination room. Once you leave, you will not be permitted to return to work or change your answers.

Practice Exam 2 Instructions

In accordance with the rules established by the California Board for Professional Engineers, Land Surveyors, and Geologists, you may use textbooks, handbooks, bound reference materials, and any approved battery- or solar-powered, silent calculator to work this examination. However, no blank papers, writing tablets, unbound scratch paper, or loose notes are permitted. Sufficient room for scratch work is provided in the Examination Booklet. You are not permitted to share or exchange materials with other examinees.

You will have $2\frac{1}{2}$ hours in which to work this examination. All questions must be worked correctly in order to receive full credit on the exam. There are no optional questions. Partial credit is not available. No credit will be given for methodology, assumptions, or work written in your Examination Booklet.

Record all of your answers on the Answer Sheet. No credit will be given for answers marked in the Examination Booklet. To simulate the flagging function on the CBT exam, the answer sheet includes a flag icon that you can use to identify the problems you want to come back to.

If you finish early, check your work and make sure that you have followed all instructions. After checking your answers, you may turn in your Examination Booklet, submit your answers, and leave the examination room. Once you leave, you will not be permitted to return to work or change your answers.

WAIT FOR PERMISSION TO BEGIN

Name: _____
        Last   First   Middle Initial

Examinee number: _____

Examination Booklet number: _____

California Civil: Seismic Principles Exam

Practice Exam 2

Practice Exam 2 Answer Sheet

56. Ⓐ Ⓑ Ⓒ Ⓓ ⚑	70. Ⓐ Ⓑ Ⓒ Ⓓ ⚑	84. Ⓐ Ⓑ Ⓒ Ⓓ ⚑	98. Ⓐ Ⓑ Ⓒ Ⓓ ⚑
57. Ⓐ Ⓑ Ⓒ Ⓓ ⚑	71. Ⓐ Ⓑ Ⓒ Ⓓ ⚑	85. Ⓐ Ⓑ Ⓒ Ⓓ ⚑	99. Ⓐ Ⓑ Ⓒ Ⓓ ⚑
58. Ⓐ Ⓑ Ⓒ Ⓓ ⚑	72. Ⓐ Ⓑ Ⓒ Ⓓ ⚑	86. Ⓐ Ⓑ Ⓒ Ⓓ ⚑	100. Ⓐ Ⓑ Ⓒ Ⓓ ⚑
59. Ⓐ Ⓑ Ⓒ Ⓓ ⚑	73. Ⓐ Ⓑ Ⓒ Ⓓ ⚑	87. Ⓐ Ⓑ Ⓒ Ⓓ ⚑	101. Ⓐ Ⓑ Ⓒ Ⓓ ⚑
60. Ⓐ Ⓑ Ⓒ Ⓓ ⚑	74. Ⓐ Ⓑ Ⓒ Ⓓ ⚑	88. Ⓐ Ⓑ Ⓒ Ⓓ ⚑	102. Ⓐ Ⓑ Ⓒ Ⓓ ⚑
61. Ⓐ Ⓑ Ⓒ Ⓓ ⚑	75. Ⓐ Ⓑ Ⓒ Ⓓ ⚑	89. Ⓐ Ⓑ Ⓒ Ⓓ ⚑	103. Ⓐ Ⓑ Ⓒ Ⓓ ⚑
62. Ⓐ Ⓑ Ⓒ Ⓓ ⚑	76. Ⓐ Ⓑ Ⓒ Ⓓ ⚑	90. Ⓐ Ⓑ Ⓒ Ⓓ ⚑	104. Ⓐ Ⓑ Ⓒ Ⓓ ⚑
63. Ⓐ Ⓑ Ⓒ Ⓓ ⚑	77. Ⓐ Ⓑ Ⓒ Ⓓ ⚑	91. Ⓐ Ⓑ Ⓒ Ⓓ ⚑	105. Ⓐ Ⓑ Ⓒ Ⓓ ⚑
64. Ⓐ Ⓑ Ⓒ Ⓓ ⚑	78. Ⓐ Ⓑ Ⓒ Ⓓ ⚑	92. Ⓐ Ⓑ Ⓒ Ⓓ ⚑	106. Ⓐ Ⓑ Ⓒ Ⓓ ⚑
65. Ⓐ Ⓑ Ⓒ Ⓓ ⚑	79. Ⓐ Ⓑ Ⓒ Ⓓ ⚑	93. Ⓐ Ⓑ Ⓒ Ⓓ ⚑	107. Ⓐ Ⓑ Ⓒ Ⓓ ⚑
66. Ⓐ Ⓑ Ⓒ Ⓓ ⚑	80. Ⓐ Ⓑ Ⓒ Ⓓ ⚑	94. Ⓐ Ⓑ Ⓒ Ⓓ ⚑	108. Ⓐ Ⓑ Ⓒ Ⓓ ⚑
67. Ⓐ Ⓑ Ⓒ Ⓓ ⚑	81. Ⓐ Ⓑ Ⓒ Ⓓ ⚑	95. Ⓐ Ⓑ Ⓒ Ⓓ ⚑	109. Ⓐ Ⓑ Ⓒ Ⓓ ⚑
68. Ⓐ Ⓑ Ⓒ Ⓓ ⚑	82. Ⓐ Ⓑ Ⓒ Ⓓ ⚑	96. Ⓐ Ⓑ Ⓒ Ⓓ ⚑	110. Ⓐ Ⓑ Ⓒ Ⓓ ⚑
69. Ⓐ Ⓑ Ⓒ Ⓓ ⚑	83. Ⓐ Ⓑ Ⓒ Ⓓ ⚑	97. Ⓐ Ⓑ Ⓒ Ⓓ ⚑	

Practice Exam 2

56. The location of an earthquake is shown. Where is the epicenter located?

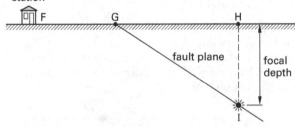

seismic network station

(A) F

(B) G

(C) H

(D) I

57. Based on ASCE/SEI7 seismic design criteria and NEHRP provisions, structures may sustain structural damage but not collapse during a

I. small earthquake

II. moderate earthquake

III. severe earthquake

(A) III only

(B) I and II only

(C) I, II, and III

(D) Structural collapse cannot be prevented in the event of an earthquake.

58. For a structure with redundancy, which statement(s) is/are correct?

I. Temporary seismic overstress may be redistributed to alternate load paths with higher resistance capacities.

II. The structure will remain laterally stable after the failure of any single element.

III. All components must remain operative for the structure to retain its lateral stability.

(A) II only

(B) III only

(C) I and II only

(D) I, II, and III

59. The structures shown are being studied. Structure I has significant physical discontinuities in its configuration and its lateral force-resisting system, including torsional irregularity. Structure II has no significant physical discontinuities in its plan and vertical configuration or in its lateral force-resisting system. $S_{DS} = 0.833$ and $S_{D1} = 0.4$, and both structures are assigned to seismic design category D. The time period is 1.2 sec for structure I and 2.0 sec for structure II. Which structure can be designed using the equivalent lateral-force procedure?

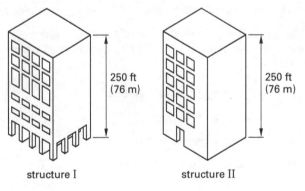

structure I   structure II

(A) I only

(B) II only

(C) I and II only

(D) neither I nor II

60. A civil engineering document should bear the seal and signature of the registered civil engineer who

    I.   prepared the document

    II. supervised the preparation of the document

    III. conveyed an engineering recommendation or decision on the document

(A) III only

(B) I and II only

(C) I and III only

(D) I, II, and III

61. Consider the two nonbuilding structures shown. Both are located on a site where $S_{DS} = 1.2$ and $S_1 = 0.6$. The monument has a fundamental period of 0.05 sec. The amusement structure has a fundamental period of 0.3 sec. Both structures are risk category II, and the weight of both structures is the same. Which structure has the smaller base shear?

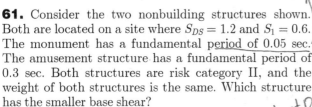

 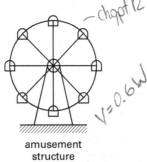

(A) the monument

(B) the amusement structure

(C) both structures have the same base shear

(D) not enough information

62. In California, which type of building construction has generally had the WORST seismic performance record?

(A) prefabricated metal buildings

(B) tilt-up construction

(C) unreinforced masonry construction

(D) moment frames with infilled walls

63. Which factors most often cause damage to the non-structural architectural elements of a building?

    I.   insufficient anchorage capacity

    II. differential seismic displacement

    III. inadequate building separations

(A) II only

(B) I and II only

(C) I and III only

(D) I, II, and III

64. A five-story police station in San Francisco with a steel eccentrically braced frame has a fundamental period of 0.6 sec. This building is constructed with a 12 ft (3.7 m) story height. Each story floor and roof has a weight of 500 kips (2200 kN).

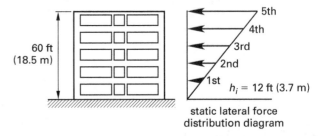

static lateral force
distribution diagram

If the distributed base shear at the fourth floor is equal to 32 kips (140 kN), the base shear would most nearly be

(A) 80 kips (360 kN)

(B) 120 kips (520 kN)

(C) 160 kips (710 kN)

(D) 200 kips (890 kN)

65. The Alquist-Priolo Earthquake Fault Zoning Act prevents the construction of which buildings on the surface trace of an active fault?

    I.   buildings used for human occupancy

    II. two-story single family wood-frame dwellings

    III. structures housing explosive and/or hazardous materials

(A) I only

(B) III only

(C) I and III only

(D) I, II, and III

66. A wood structural panel roof diaphragm is shown in plan view. At the diaphragm boundaries, because of openings (e.g., windows or doors), the shear walls are not the full width of the building. For north-south loading, which members transmit unsupported horizontal diaphragm shear to the supporting shear walls?

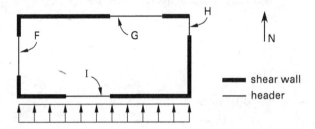

(A)   I only

(B)   F and H only

(C)   I and G only

(D)   F, G, H, and I

67. A force is acting at the top of a frame as shown. The supporting columns are of equal height. The left column is pinned, and the right column is fixed at the base. The modulus of elasticity, E, and moment of inertia, I, for both columns are the same. The top plate is rigid.

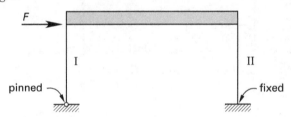

Which relationship represents the resisting shears for each column?

(A) $V_I = \frac{1}{8} V_{II}$

(B) $V_I = \frac{1}{4} V_{II}$

(C) $V_I = 4 V_{II}$

(D) $V_I = 16 V_{II}$

68. Which of the steel frame structures shown are concentrically braced frames?

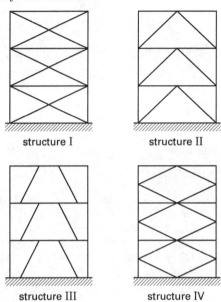

(A)   III only

(B)   I and II only

(C)   IV only

(D)   I, II, and IV only

69. According to ASCE/SEI7, which can be idealized as a rigid diaphragm?

   I.   concrete-filled metal deck with a span-to-depth ratio of three and no horizontal irregularities

  II.   wood structural panels with steel and concrete braced framed vertical elements

III.   untopped steel decks in a single-family residential building of light-frame construction

(A)   I only

(B)   III only

(C)   I and II only

(D)   II and III only

70. A magnitude increase of one unit on the Richter scale represents how much of an increase in measured maximum amplitude on a seismometer trace?

(A)   1 time

(B)   10 times

(C)   32 times

(D)   100 times

71. Which structures can be designed and detailed by a professional civil engineer?

    I.   public schools

    II. high-rise office buildings

    III. vehicle bridge structures

(A) II only

(B) III only

(C) I and II only

(D) II and III only

72. The tank shown contains fire-suppression material required for the protection of the building. The tank is mounted on structural steel legs, is well braced, and has adequate anchorage to the building. It is constructed of high-deformability materials. The weight of the tank when full is W and $S_{DS} = 1.2$.

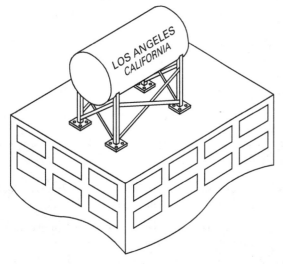

The calculated lateral force on this tank is most nearly

(A) $0.50\,W$

(B) $0.60\,W$

(C) $0.90\,W$

(D) $3.00\,W$

73. What type of steel frame system is appropriate for a 150 ft tall, twelve-story office building assigned to seismic design category D?

    I.   ordinary concentrically braced frame $\times$

    II. ordinary moment-resisting frame $\vee$

    III. eccentrically braced frame

(A) II only

(B) III only

(C) I and III only

(D) I, II, and III

74. A building is sketched in plan view as shown. The building is stable. The rigidities of the north, east, and west walls are the same. The center of mass (CM) and center of rigidity (CR) are given.

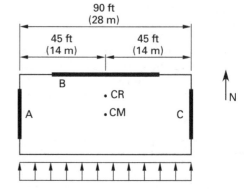

Under north-south loading, if accidental eccentricity is neglected, which statement is correct?

(A) The eccentricity causes a torsional moment.

(B) There is no torsion.

(C) The earthquake shear force is resisted directly by wall B.

(D) Wall A and wall C have different amounts of displacement.

75. The story (floor) deflections from the frame analysis of a four-story commercial building are illustrated as shown.

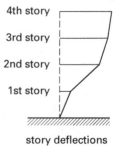

story deflections

What kind of vertical structural irregularity can be associated with this building?

(A) weak story

(B) soft story

(C) mass (weight) irregularity

(D) vertical geometric irregularity

76. The wood structural panel roof diaphragm of a one-story residential building is shown in plan view. The lateral force on the roof diaphragm is 250 lbf/ft (3600 N/m).

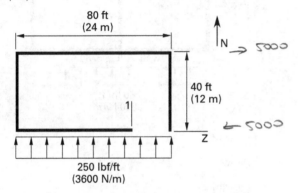

80 ft
(24 m)

40 ft
(12 m)

250 lbf/ft
(3600 N/m)

If the force in the chord member at the intersection of lines Z and 1 is 3750 lbf (16 500 N), the length of the south wall is most nearly

(A) 40 ft (12 m)

(B) 55 ft (17 m)

(C) 60 ft (18 m)

(D) 75 ft (23 m)

77. The spectral response acceleration for a site-specific maximum considered earthquake (MCE_R) is

(A) equal to the probabilistic MCE_R

(B) equal to the deterministic MCE_R

(C) the lesser of the probabilistic and the deterministic MCE_Rs

(D) the larger of the probabilistic and the deterministic MCE_Rs

78. A one-story warehouse with a concrete roof diaphragm and shear walls is shown. The roof and shear walls have a uniform thickness. The roof and wall dead loads are 25 lbf/ft² (1200 N/m²) and 45 lbf/ft² (2200 N/m²), respectively. The tabulated rigidities, R_{tab}, of the walls are as shown. The left corner, point O, is the origin.

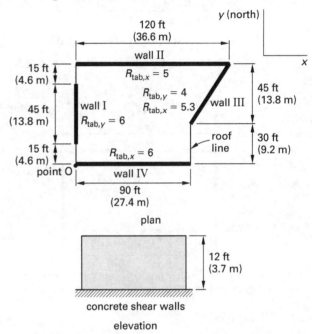

plan

12 ft
(3.7 m)

concrete shear walls

elevation

Using the full wall height for the building center of mass calculation, the locations of the building center of mass and the building center of rigidity in the x-direction are most nearly

(A) 42 ft and 32 ft (13 m and 10 m)

(B) 47 ft and 37 ft (14 m and 11 m)

(C) 52 ft and 42 ft (16 m and 13 m)

(D) 57 ft and 47 ft (17 m and 14 m)

79. Which of the given statements is/are true regarding wood structural panel shear walls that are part of a one-story wood frame structure?

  I. They are known as resisting elements.

  II. They distribute the lateral force to the roof diaphragm.

  III. They resist torsional shear.

(A) I only

(B) III only

(C) I and III only

(D) I, II, and III

80. The plan view of a one-story masonry shear wall structure with a rigid roof diaphragm is shown. The locations of the centers of mass (CM) and rigidity (CR) are given.

	center of mass		center of rigidity	
$\bar{x}$	32 ft	(9.8 m)	45 ft	(13.7 m)
$\bar{y}$	25 ft	(7.6 m)	15 ft	(4.6 m)

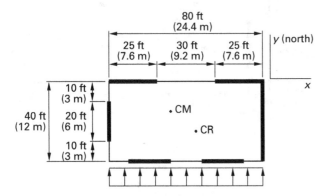

For north-south loading, what is the design eccentricity to be used in calculating the torsional design moment?

(A)   10 ft (3.0 m)

(B)   13 ft (4.0 m)

(C)   17 ft (5.1 m)

(D)   19 ft (5.8 m)

81. Which of the criteria shown would a group of equally qualified engineers consider to evaluate whether a professional civil engineer is in responsible charge of his or her professional engineering work?

   I.   Does the engineer demonstrate that he or she personally made engineering decisions?

  II.   Does the engineer possess sufficient knowledge of the project?

 III.   Does the engineer defend decisions in an adversarial situation?

(A)   I only

(B)   I and II only

(C)   II and III only

(D)   I, II, and III

82. The concrete walls of a stable one-story building with a rigid roof diaphragm are 15 ft (5 m) in height. The tabulated rigidities, R_{tab}, of the walls are as shown. Only the walls are designed to resist lateral force in the north-south direction. Ignoring the effects of torsion and eccentricity, which wall resists the smallest proportion of the lateral force shown?

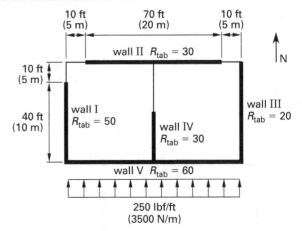

(A)   wall II

(B)   wall III

(C)   wall IV

(D)   all walls resist equally

83. Which statement is NOT correct?

(A)   California's earthquakes are relatively shallow.

(B)   The San Andreas fault movement is left-lateral.

(C)   In California, many smaller faults branch from and join the San Andreas fault.

(D)   The Loma Prieta earthquake of 1989 originated on the San Andreas fault.

84. For a two-story, single-family residential building of wood-frame construction, which registered professionals can sign final calculations of dead plus live plus seismic combinations of loads?

(A)   architects

(B)   civil engineers

(C)   structural engineers

(D)   all of the above

85. A mass of 25,000 lbf (11 000 kg) is supported on two vertical members with a lateral stiffness of 15,000 lbf/in (2.7×10^6 N/m) each. The columns have no mass, are of equal height, and are fixed at both ends. The natural period, T, for the frame shown is 0.3 sec.

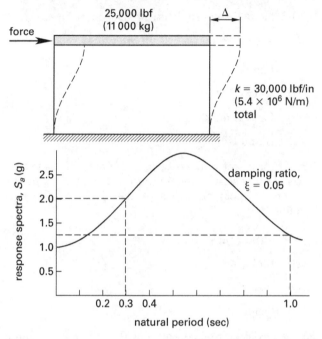

What is most nearly the deflection, Δ, due to seismic base shear?

(A)   0.8 in (20 mm)

(B)   1.7 in (40 mm)

(C)   2.5 in (64 mm)

(D)   3.3 in (84 mm)

86. The east wall of a one-story building is shown. The masonry shear wall, steel frame, and concrete shear wall are fixed on both the top and bottom. The absolute rigidity of this wall is 15 kips/in (40 kN/mm). The lateral loads carried by the masonry shear wall, steel frame, and concrete shear wall are 40 kips (180 kN), 15 kips (70 kN), and 20 kips (90 kN), respectively.

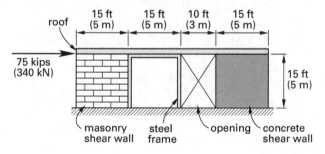

Which statement(s) is/are FALSE?

I.   The steel frame has the largest relative rigidity.

II.   The masonry shear wall has the largest absolute rigidity.

III.   The relative rigidities are equal.

IV.   The deflections of the shear walls and steel frame are equal.

(A)   I only

(B)   I and III only

(C)   III and IV only

(D)   I, II, and IV only

87. A typical highway bridge structure is shown. This structure consists of a reinforced concrete slab super-structure and multicolumn bents. The columns are fixed at both top and bottom. The abutments can adequately resist earthquake forces, while the columns lack adequate confinement and shear capacity.

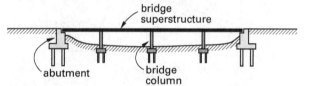

What options are available to the engineers to retrofit the concrete columns of this bridge structure?

I.   full-height column casing retrofit

II.   partial-height column casing retrofit

III.   seismic anchor slab

(A)   III only

(B)   I and II only

(C)   I and III only

(D)   I, II, and III

88. Which statement(s) about a moment-resisting frame is/are true?

I.   It carries both vertical and lateral loads.

II.   It has partially rigid joints.

III.   It resists forces by flexure.

(A)   I only

(B)   III only

(C)   I and III only

(D)   I, II, and III

89. The wood structural panel shear walls of a one-story wood-frame residential building is shown.

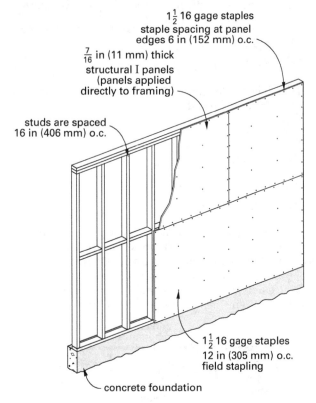

$1\frac{1}{2}$ 16 gage staples
staple spacing at panel
edges 6 in (152 mm) o.c.

$\frac{7}{16}$ in (11 mm) thick
structural I panels
(panels applied
directly to framing)

studs are spaced
16 in (406 mm) o.c.

$1\frac{1}{2}$ 16 gage staples
12 in (305 mm) o.c.
field stapling

concrete foundation

Using the appropriate IBC table, what is most nearly the maximum allowable shear in the wood structural panel shear walls?

(A)   110 lbf/ft (1605 N/m)

(B)   155 lbf/ft (2260 N/m)

(C)   170 lbf/ft (2480 N/m)

(D)   185 lbf/ft (2700 N/m)

90. Site class F includes

   I.   soils that are liquefiable

   II.   soils with more than 10 ft (3 m) of soft clay with the plasticity index, PI, greater than or equal to 20, and the moisture content, w_{mc}, greater than or equal to 40%

   III.   soils with more than 25 ft (7.6 m) of soft clay with the plasticity index, PI, greater than 75

(A)   I only

(B)   III only

(C)   I and III only

(D)   I, II, and III

91. A five-story building is risk category IV and seismic design category D and has a type 1a horizontal irregularity. The building can be analyzed using the

   I.   equivalent lateral-force analysis procedure

   II.   modal response spectrum analysis

   III.   seismic response history procedure

(A)   I only

(B)   II only

(C)   I and II only

(D)   II and III only

92. Typical retrofit costs per square foot (square meter) for the seismic rehabilitation of an existing building depend most on which of the given factors of a building?

(A)   structural system

(B)   roof type

(C)   material

(D)   height

93. An existing wood-frame single-family home is built on the side of a steep hill. A portion of the house is set on bare wood posts (posts-and-piers foundation system) as shown. The piers are deep enough in the ground and rest on a solid base.

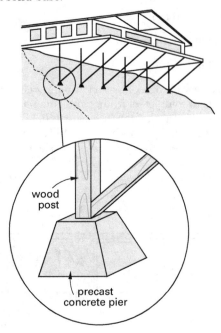

wood
post

precast
concrete pier

What strengthening technique will improve the seismic resistance of this structure?

    I.   reinforcing the top of the post at its connection to the beam

    II. reinforcing the post-and-pier connections

    III. reinforcing the posts with X-diagonal braces

(A) II only

(B) III only

(C) II and III only

(D) I, II, and III

94. Which ASCE/SEI7 provision states the following: All structural framing elements, and their connections, not required by design to be part of the lateral-force resisting system, need to be adequate to maintain support of design dead plus live loads when subjected to the expected deformations caused by seismic forces?

(A) building separations

(B) ties and continuity

(C) deformation compatibility

(D) *P*-delta effects

95. A 220 ft (67 m) high-rise residential building with a seismic design category of E is being constructed in downtown Los Angeles. According to ASCE/SEI7 design provisions, a special reinforced concrete shear wall may be used as part of the lateral force-resisting system as long as

    I.   the structure does not have a type 1b horizontal structural irregularity

    II. the shear walls in any one plane resist no more than 60% of the total seismic forces in each direction, neglecting accidental torsional effects

    III. dual systems with special concrete moment frames are capable of resisting at least 25% of the prescribed seismic forces

(A) I only

(B) III only

(C) I and II only

(D) I, II, and III

96. Consider a building with a flexible diaphragm. Additional new braced frames (vertical resisting elements) are needed to resist seismic forces because the existing elements are overstressed. Which of the alternatives shown will reduce seismic demand on the bracing elements?

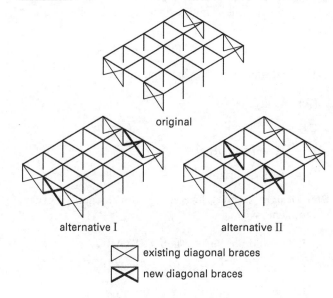

original

alternative I   alternative II

⊠ existing diagonal braces

◼⊠ new diagonal braces

(A) I only

(B) II only

(C) both I and II

(D) neither I nor II

97. After an earthquake disaster strikes a community, who can perform building safety evaluations and damage inspections if requested by the local building officials?

    I.   qualified building inspectors

    II. civil engineers

    III. structural engineers

(A) I only

(B) III only

(C) II and III only

(D) I, II, and III

98. A geotechnical engineering report finds an existing foundation with deficient piles/drilled piers. The capacity of the piles or drilled-piers foundation can be increased by

    I.   removing the existing pile caps, driving additional piles, and providing new pile caps of larger size

    II.   reducing loads on the piles or piers

    III.   adding tie beams between pile caps

(A) I only

(B) I and II only

(C) II and III only

(D) I, II, and III

99. Design seismic forces, F_p, on a nonstructural component

    I.   should not be less than $0.3S_{DS}I_pW_p$

    II.   should not be less than $1.6S_{DS}I_pW_p$

    III.   need not be more than $1.6S_{DS}I_pW_p$

(A) I only

(B) III only

(C) I and III only

(D) II and III only

100. According to ASCE/SEI7 requirements, if all mechanical components have an I_p value of 1.0, which do NOT require seismic supports?

    I.   HVAC ducts in a hotel that are suspended from hangers 6 in (153 mm) long

    II.   HVAC ducts in a hotel whose cross sections are 2 ft by 2 ft (60 cm by 60 cm) and are installed in a configuration where there will be no impact with larger ducts

    III.   HVAC fans weighing 90 lbf (400 N) each that are installed in-line with the ductwork

(A) I only

(B) II only

(C) III only

(D) I and II only

101. Which deficiencies cause a structure to lack a complete load path for seismic force effects?

(A) a discontinuous chord

(B) a missing collector

(C) a deficient connection

(D) all of the above

102. What is the approximate fundamental period for a 50 ft (15 m) structure whose steel moment-resisting frame resists 100% of the required seismic forces and is adjoined by less rigid components?

(A) 0.3 sec

(B) 0.4 sec

(C) 0.5 sec

(D) 0.6 sec

103. Which is true of chord and collector elements?

(A) A structural element cannot serve as both a chord and a collector.

(B) Chords are generally perpendicular to lateral loads, and collectors are generally parallel to lateral loads.

(C) Chords carry only compression forces, and collectors carry only tension forces.

(D) Chords carry only shear forces, and collectors carry only compression forces.

104. According to ASCE/SEI7, which of the following does NOT need to be included when calculating a structure's effective seismic weight?

(A) dead loads

(B) floor live loads in storage areas

(C) floor live loads in public garages

(D) flat roof snow loads

105. A 150 ft (46 m) tall structure is in seismic design category D. Which structural system may NOT be used?

(A) special steel moment-resisting frame

(B) ordinary reinforced concrete shear wall paired with special moment-resisting frame

(C) steel eccentrically braced frame

(D) steel and concrete composite with concentrically braced frames

106. According to ASCE/SEI7, which of the following analytical procedures may be used for a 150 ft structure of light frame construction with a reentrant corner irregularity?

   I.   equivalent lateral force analysis

   II. modal response spectrum analysis

   III. seismic response history procedures

   (A) I only

   (B) II only

   (C) I and II only

   (D) I, II, and III

107. Which of the following must be considered when determining a building's redundancy factor?

   I.   the spatial distribution and weight of each floor

   II. the location of lateral frames on the perimeter of the building

   III. the effect of removing portions of the frame

   (A) I only

   (B) III only

   (C) I and II only

   (D) II and III only

108. A special moment-resisting steel frame structure has an importance factor of 1.0. The structural analysis finds that the elastic deflection of level 4 is 2.0 in (51 mm), and the elastic deflection of level 5 is 2.8 in (71 mm). The design story drift for level 5 of the structure is most nearly

   (A) 0.8 in (20 mm)

   (B) 2.8 in (71 mm)

   (C) 4.4 in (110 mm)

   (D) 15 in (380 mm)

109. According to the IBC, which type of reinforced concrete work does NOT require continuous inspection?

   (A) tensioning the prestressing tendons

   (B) grouting the prestressing strands

   (C) placing the reinforcing steel

   (D) sampling fresh concrete for strength tests

110. A structure is in risk category III, and its site class is D. S_{DS} is 0.40, and S_{D1} is 0.10. The seismic design category for the structure is

   (A) A

   (B) B

   (C) C

   (D) D

STOP!

DO NOT CONTINUE!

This concludes the California Civil: Seismic Principles Exam. If you finish early, check your work and make sure that you have followed all instructions. After checking your answers, you may turn in your examination booklet and answer sheet and leave the examination room. Once you leave, you will not be permitted to return to work or change your answers.

Answer Keys

Practice Exam 1 Answer Key

#	Ans	#	Ans	#	Ans	#	Ans
1.	B	15.	B	29.	A	43.	C
2.	A	16.	D	30.	D	44.	D
3.	B	17.	D	31.	B	45.	D
4.	A	18.	A	32.	C	46.	D
5.	A	19.	C	33.	A	47.	C
6.	D	20.	C	34.	C	48.	B
7.	B	21.	C	35.	A	49.	B
8.	D	22.	A	36.	D	50.	D
9.	D	23.	A	37.	D	51.	D
10.	D	24.	B	38.	A	52.	A
11.	C	25.	C	39.	D	53.	C
12.	D	26.	A	40.	C	54.	D
13.	B	27.	B	41.	B	55.	C
14.	B	28.	B	42.	B		

Practice Exam 2 Answer Key

#	Ans	#	Ans	#	Ans	#	Ans
56.	C	70.	B	84.	D	98.	D
57.	A	71.	D	85.	B	99.	C
58.	C	72.	C	86.	B	100.	D
59.	D	73.	B	87.	B	101.	D
60.	D	74.	B	88.	C	102.	D
61.	A	75.	B	89.	C	103.	B
62.	C	76.	C	90.	C	104.	C
63.	B	77.	C	91.	D	105.	B
64.	B	78.	C	92.	A	106.	D
65.	A	79.	A	93.	D	107.	D
66.	B	80.	C	94.	C	108.	C
67.	B	81.	B	95.	C	109.	C
68.	D	82.	B	96.	C	110.	C
69.	A	83.	B	97.	D		

Solutions
Practice Exam 1

1. The Alquist-Priolo Earthquake Fault Zoning Act was passed in 1972 as a direct result of the 1971 San Fernando earthquake. This earthquake was characterized by extensive surface fault ruptures that damaged numerous residential homes, commercial buildings, and other structures.

The Alquist-Priolo Earthquake Fault Zoning Act is intended to prevent the construction of buildings utilized for human occupancy on the surface trace of active faults. The provisions of this law require that buildings for human occupancy must be at least 50 ft (15.24 m) away from an active fault trace. Prior to January 1, 1994, Earthquake Fault Zones were called Special Studies Zones.

This act requires the State Geologist to establish regulatory zones around the surface traces of active faults and to issue the appropriate maps. Zone maps are available from the California Division of Mines and Geology, or they may be viewed at city or county planning departments and some real estate offices.

The answer is (B).

2. According to ASCE/SEI7 Sec. 11.4.5, the earthquake design of a building is $\frac{2}{3}$ the maximum considered earthquake design motion. The maximum considered earthquake coefficients can be obtained from ASCE/SEI7 Fig. 22-1 through Fig. 22-6.

The answer is (A).

3.

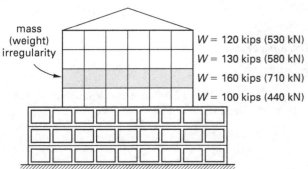

mass (weight) irregularity

$W = 120$ kips (530 kN)
$W = 130$ kips (580 kN)
$W = 160$ kips (710 kN)
$W = 100$ kips (440 kN)

ASCE/SEI7 Table 12.3-2 lists and defines the five types of vertical structural irregularities: soft-story irregularity, mass (weight) irregularity, vertical geometric irregularity, in-plane discontinuity irregularity, and

discontinuity in lateral strength irregularity. Of these, the structure shown has a mass (weight) irregularity in the second story because the mass is more than 150% of the effective mass of the story below.

The answer is (B).

4. *SI Solution*

The mapped acceleration parameters are given in the problem statement as 0.2 and 0.5. The site class is not given, therefore, it is assumed to be site class D according to ASCE/SEI7 Sec. 11.4.3.

To obtain the site coefficients and the risk-targeted maximum considered earthquake (MCE$_R$) spectral response acceleration parameters, use ASCE/SEI7 Sec. 11.4.4. For $S_S = 0.5$ and site class D, the site coefficient, F_a, is 1.4 [ASCE/SEI7 Table 11.4-1]. For $S_1 = 0.2$ and site class D, the site coefficient, F_v, is 2.2 [ASCE/SEI7 Table 11.4-2].

The MCE$_R$ spectral response acceleration for short periods, S_{MS}, and a 1 s period, S_{M1}, is given by ASCE/SEI7 Eq. 11.4-1 and Eq. 11.4-2.

$$S_{MS} = F_a S_S = (1.4)(0.5) = 0.7$$
$$S_{M1} = F_v S_1 = (2.2)(0.2) = 0.44$$

The design spectral acceleration parameters are given by ASCE/SEI7 Eq. 11.4-3 and Eq. 11.4-4.

$$S_{DS} = \frac{2}{3} S_{MS} = \left(\frac{2}{3}\right)(0.7) = 0.47$$

$$S_{D1} = \frac{2}{3} S_{M1} = \left(\frac{2}{3}\right)(0.44) = 0.29$$

Determine the risk category for the building based on ASCE/SEI7 Table 1.5-1. For this building, the risk category is IV.

From ASCE/SEI7 Sec. 11.5.1 and Table 1.5-2, the importance factor, I_e, is 1.5. From ASCE/SEI7 Table 12.2-1, the response modification factor, R, is 8.

Determine the building's time period based on ASCE/SEI7 Sec. 12.8.2. From ASCE/SEI7 Table 12.8-2, $C_t = 0.0731$ and $x = 0.75$.

$$T_a = C_t h_n^x = (0.0731)(47 \text{ m})^{0.75} = 1.312 \text{ s}$$

ASCE/SEI7 Sec. 12.8 states the procedure for the equivalent lateral force method.

$$C_s = \frac{S_{DS}}{\dfrac{R}{I_e}} = \frac{0.47}{\dfrac{8}{1.5}}$$

$$= 0.088 \quad \text{[ASCE/SEI7 Eq. 12.8-2]}$$

$$C_{s,\max} = \frac{S_{D1}}{T_a\left(\dfrac{R}{I_e}\right)} = \frac{0.29}{(1.312 \text{ s})\left(\dfrac{8}{1.5}\right)}$$

$$= 0.42 \quad \text{[ASCE/SEI7 Eq. 12.8-3]}$$

ASCE/SEI7 Eq. 12.8-3 is valid for $T \quad T_L$. (In California, T_L is either 8 s or 12 s, so ASCE/SEI7 Eq. 12.8-4 will not be valid.)

$$C_s = 0.044 S_{DS} I_e \geq 0.01 \quad \text{[ASCE/SEI7 Eq. 12.8-5]}$$

$$= (0.044)(0.47)(1.5)$$

$$= 0.031 \quad [\geq 0.01, \text{OK}]$$

$$C_{s,\text{gov}} = 0.031$$

Per ASCE/SEI7 Eq. 12.8-1, the base shear is

$$V = C_s W = 0.42 W$$

The answer is (A).

Customary U.S. Solution

The mapped acceleration parameters are given in the problem statement as 0.2 and 0.5. The site class is not given, therefore, it is assumed to be site class D according to ASCE/SEI7 Sec. 11.4.3.

To obtain the site coefficients and the risk-targeted maximum considered earthquake (MCE_R) spectral response acceleration parameters, use ASCE/SEI7 Sec. 11.4.4. For $S_S = 0.5$ and site class D, the site coefficient, F_a, is 1.4 [ASCE/SEI7 Table 11.4-1]. For $S_1 = 0.2$ and site class D, the site coefficient, F_v, is 2.2 [ASCE/SEI7 Table 11.4-2].

The MCE_R spectral response acceleration for short periods, S_{MS}, and a 1 sec period, S_{M1}, is given by ASCE/SEI7 Eq. 11.4-1 and Eq. 11.4-2.

$$S_{MS} = F_a S_S = (1.4)(0.5) = 0.7$$

$$S_{M1} = F_v S_1 = (2.2)(0.44) = 0.44$$

The design spectral acceleration parameters are given by ASCE/SEI7 Eq. 11.4-3 and Eq. 11.4-4.

$$S_{DS} = \tfrac{2}{3} S_{MS} = \left(\tfrac{2}{3}\right)(0.7) = 0.47$$

$$S_{D1} = \tfrac{2}{3} S_{M1} = \left(\tfrac{2}{3}\right)(0.44) = 0.29$$

Determine the risk category for the building based on ASCE/SEI7 Table 1.5-1. For this building, the risk category is IV.

From ASCE/SEI7 Sec. 11.5.1 and Table 1.5-2, the importance factor, I_e, is 1.5. From ASCE/SEI7 Table 12.2-1, the response modification factor, R, is 8.

Determine the time period of the building based on ASCE/SEI7 Sec. 12.8.2. From ASCE/SEI7 Table 12.8-2, $C_t = 0.03$ and $x = 0.75$.

$$T_a = C_t h_n^x = (0.03)(155 \text{ ft})^{0.75} = 1.318 \text{ sec}$$

ASCE/SEI7 Sec. 12.8 states the procedure for the equivalent lateral force method.

$$C_s = \frac{S_{DS}}{\dfrac{R}{I_e}} = \frac{0.47}{\dfrac{8}{1.5}}$$

$$= 0.088 \quad \text{[ASCE/SEI7 Eq. 12.8-2]}$$

$$C_{s,\max} = \frac{S_{D1}}{T_a\left(\dfrac{R}{I_e}\right)} = \frac{0.29}{(1.318 \text{ sec})\left(\dfrac{8}{1.5}\right)}$$

$$= 0.42 \quad \text{[ASCE/SEI7 Eq. 12.8-3]}$$

ASCE/SEI7 Eq. 12.8-3 is valid for $T \quad T_L$. (In California, T_L is either 8 sec or 12 sec, so ASCE/SEI7 Eq. 12.8-4 will not be valid.)

$$C_{s,\min} = 0.044 S_{DS} I_e \geq 0.01 \quad \text{[ASCE/SEI7 Eq. 12.8-5]}$$

$$= (0.044)(0.47)(1.5)$$

$$= 0.031 \quad [\geq 0.01, \text{OK}]$$

$$C_{s,\text{gov}} = 0.031$$

Per ASCE/SEI7 Eq. 12.8-1, the base shear is

$$V = C_s W = 0.42 W$$

The answer is (A).

5. *SI Solution*

Use ASCE/SEI7 Sec. 12.8.3 and the equations shown to determine the distributed shear at level 3.

$$F_x = C_{vx} V \quad \text{[ASCE/SEI7 Eq. 12.8-11]}$$

$$C_{vx} = \frac{w_x h_x^k}{\displaystyle\sum_{i=1}^{n} w_i h_i^k} \quad \text{[ASCE/SEI7 Eq. 12.8-12]}$$

From the problem statement, V is 220 000 N. The story height and weight are equal. Since the height of the structure is given as 24 m, each story height is 3 m. At each story, weight is w. h_x is the height from the base to level x. Since the fundamental period is 0.68 s, the value of k should be interpolated. For structures having a period of 0.5 s or less, $k = 1$, and for structures having a period greater than 2.5 s, $k = 2$.

$$k_{T = 0.68 \text{ s}} = 1 + \left(\frac{0.68 \text{ s} - 0.5 \text{ s}}{2.5 \text{ s} - 0.5 \text{ s}}\right)(2 - 1) = 1.09$$

A table is the easiest way to set up the data for calculating and recording story shears.

level x	h_x (m)	w_x (N)	$w_x h_x^{\,k}$	$\dfrac{w_x h_x^{\,k}}{\sum w_x h_x^{\,k}}$	F_x (N)
8	24	w	$31.95w$	0.23	50 600
7	21	w	$27.62w$	0.20	44 000
6	18	w	$23.35w$	0.17	37 400
5	15	w	$19.14w$	0.14	30 800
4	12	w	$15.01w$	0.11	24 200
3	**9**	w	**$10.97w$**	**0.00**	**17 600**
2	6	w	$7.05w$	0.051	11 000
1	3	w	$3.31w$	0.024	4400
			$\sum w_x h_x^{\,k}$ $138.40\,w$		

The distributed base shear at level three is 17 600 N (18 000 N).

The answer is (A).

Customary U.S. Solution

Use ASCE/SEI7 Sec. 12.8.3 and the equations shown to determine the distributed shear at level 3.

$$F_x = C_{vx}V \qquad [\text{ASCE/SEI7 Eq. 12.8-11}]$$

$$C_{vx} = \frac{w_x h_x^{\,k}}{\displaystyle\sum_{i\,=1}^{n} w_i h_i^{\,k}} \qquad [\text{ASCE/SEI7 Eq. 12.8-12}]$$

From the problem statement, V is 49,500 lbf. The story height and weight are equal. Since the height of the structure is given as 80 ft, each story height is 10 ft. At each story, weight is w. h_x is the height from the base to level x. Since the fundamental period is 0.68 sec, the value of k should be interpolated. For structures having a period of 0.5 sec or less, $k = 1$, and for structures having a period greater than 2.5 sec, $k = 2$.

$$k_{T = 0.68 \text{ sec}} = 1 + \left(\frac{0.68 \text{ sec} - 0.5 \text{ sec}}{2.5 \text{ sec} - 0.5 \text{ sec}}\right)(2 - 1) = 1.09$$

A table is the easiest way to set up the data for calculating and recording story shears.

level x	h_x (ft)	w_x (lbf)	$w_x h_x^{\,k}$	$\dfrac{w_x h_x^{\,k}}{\sum w_x h_x^{\,k}}$	F_x (lbf)
8	80	w	$118.68w$	0.23	11,385
7	70	w	$102.60w$	0.20	9900
6	60	w	$86.73w$	0.17	8415
5	50	w	$71.10w$	0.14	6930
4	40	w	$55.75w$	0.11	5445
3	**30**	**w**	**$40.74w$**	**0.00**	**3960**
2	20	w	$26.19w$	0.051	2475
1	10	w	$12.30w$	0.024	990
			$\sum w_x h_x^{\,k}$ $514.10w$		

The distributed base shear at level three is 3960 lbf (4000 lbf).

The answer is (A).

6. There are geologic and site conditions that can lead to building structural damage and threaten life in the event of an earthquake.

Liquefaction can result in large settlement or lateral spreading during or immediately after an earthquake. *Liquefaction* is a phenomenon in soils in which the shear strength of soils is drastically reduced due to the sudden application of an earthquake force. The soil behaves like a liquid as a result of an increase in pore water pressure. It occurs most often in soils that are saturated in a loose condition and are made of cohesionless particles, such as sand. Liquefaction may result in vertical settlement as the soil squeezes water out and densifies after the event. This can cause a loss of foundation support for spread footings or damage to utilities. The lateral spreading of liquefied soils occurs in sloping areas and can be destructive to foundation systems, roads, or bridge abutments.

Many fault movements occur deep in the earth and do not have surface expressions other than distortion of the ground as shown. Some faults propagate up to the surface and are seen as an actual cracking or breaking of the ground surface along the fault during an earthquake. It is not economically feasible to build structures to withstand such large displacements because a structure constructed over an active fault can be severely damaged if the ground ruptures.

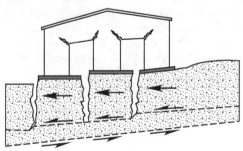

Slope failure causes foundation damage and/or loss of foundation support. Buildings may be damaged if they are located in the path of debris flows or rock falls. Seismically induced slope failures have been observed in nonliquefaction-susceptible soils, rock, slopes, and compacted fill slopes.

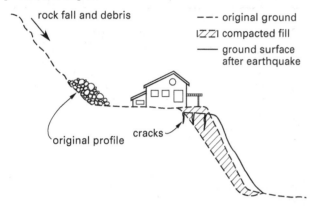

The answer is (D).

7. Use ASCE/SEI7 Sec. 13.3 to determine the seismic demands on nonstructural components. ASCE/SEI7 Eq. 13.3-1 gives a component's seismic design force, F_p, to be applied to the center of gravity.

$$F_p = \left(\frac{0.4a_p S_{DS} W_p}{\dfrac{R_p}{I_p}} \right) \left(1 + 2\frac{z}{h} \right)$$

[ASCE/SEI 7 Eq. 13.3-1]

$$F_p = \left(\frac{0.4a_p S_{DS} W_p}{\dfrac{R_p}{I_p}} \right) \left(1 + 2\frac{z}{h} \right)$$

[ASCE/SEI 7 Eq. 13.3-1]

F_p need not be greater than $F_p = 1.6S_{DS}I_p W_p$, nor less than $F_p = 0.3S_{DS}I_p W_p$ [ASCE/SEI7 Eq. 13.3-2 and Eq. 13.3-3]. a_p, the component amplification factor, is selected from ASCE/SEI7 Table 13.5-1 or Table 13.6-1 and varies from 1.0 to 2.5. R_p, the component response modification factor, is also selected from ASCE/SEI7 Table 13.5-1 or Table 13.6-1 and varies from 1.0 to 12. z is the height in structure of the point of attachment of component with respect to the base, and h is the average roof height with respect to the base. S_{DS} is determined based on ASCE/SEI7 Sec. 11.4.5.

The answer is (B).

8. Unreinforced masonry (URM) buildings, particularly bearing wall types, are considered the most hazardous form of construction not only in California, but also in other parts of the country and abroad. URM construction is no longer allowed in California, but many older URM buildings still exist. Buildings made of reinforced masonry have performed better.

Steel frame buildings have a good performance record, particularly in avoiding collapse. However, they are not damage-free during strong shakings because they absorb energy and deform. For braced steel frames, failed or buckled braces have been observed. For moment-frames, damage to primary members and distress at moment connections have been observed.

Prior to the use of ductile concrete (approximately the mid-1970s), concrete frame structures were made of nonductile concrete. Concrete structures are heavier and less yielding than similar steel structures. Past severe damage and collapse have been linked to design of joints and connections that lacked the necessary strength and ductility to withstand damage and separation forces.

Steel and concrete frame buildings with unreinforced masonry infill walls possess the additional hazard that the masonry walls can crack and fall on occupants or passersby.

The answer is (D).

9. *SI Solution*

ASCE/SEI7 Sec. 12.6 and Table 12.6-1 specify how to select an appropriate lateral-force procedure for the design of seismic force-resisting systems. The table lists three possible design procedures: the equivalent lateral-force procedure [ASCE/SEI7 Sec. 12.8], the modal response spectrum analysis [ASCE/SEI7 Sec. 12.9], and the seismic response history procedure [ASCE/SEI7 Chap. 16]. Which procedures are permitted or not permitted for a given structure depends on the seismic design category of the structure, the risk category of the structure, irregularities in the structure, and the fundamental period of the structure.

All three structures are seismic design category F and risk category III, so the only unknown factor that could determine which analytical procedures are permitted for each structure are the fundamental period and the fundamental period over a short period.

From ASCE/SEI7 Sec. 11.4.6, because S_{D1} and S_{DS} are the same for all three structures, the fundamental period over a short period is the same for all three structures. The fundamental period over a short period is

$$T_s = \frac{S_{D1}}{S_{DS}} = \frac{0.6}{0.75} = 0.8 \text{ s}$$

Per ASCE/SEI7 Table 12.6-1, for a structure of seismic design category F and risk category III to qualify for the equivalent lateral-force procedure, the fundamental period must be less than

$$3.5\,T_s = (3.5)(0.8\text{ s}) = 2.8\text{ s}$$

The approximate fundamental period of the structures can be found using ASCE/SEI7 Eq. 12.8-7. From ASCE/SEI7 Table 12.8-2, the coefficient C_t for a steel moment-resisting frame is 0.0724, and the coefficient x for a steel moment-resisting frame is 0.8. The approximate fundamental period of structure I is

$$T_a = C_t h_n^x = (0.0724)(18\text{ m})^{0.8} = 0.731\text{ s}$$

0.731 s < 2.8 s, so structure I can be analyzed using the equivalent lateral-force procedure.

The approximate fundamental period of structure II is

$$T_a = C_t h_n^x = (0.0724)(49\text{ m})^{0.8}$$
$$= 1.63\text{ s}$$

1.63 s < 2.8 s, so structure II can be analyzed using the equivalent lateral-force procedure.

The approximate fundamental period of structure III is

$$T_a = C_t h_n^x = (0.0724)(76\text{ m})^{0.8}$$
$$= 2.31\text{ s}$$

2.31 s < 2.8 s, so structure III can be analyzed using the equivalent lateral-force procedure.

Customary U.S. Solution

ASCE/SEI7 Sec. 12.6 and Table 12.6-1 specify how to select an appropriate lateral-force procedure for the design of seismic force-resisting systems. The table lists three possible design procedures: the equivalent lateral-force procedure [ASCE/SEI7 Sec. 12.8], the modal response spectrum analysis [ASCE/SEI7 Sec. 12.9], and the seismic response history procedure [ASCE/SEI7 Chap. 16]. Which procedures are permitted or not permitted for a given structure depends on the seismic design category of the structure, the risk category of the structure, irregularities in the structure, and the fundamental period of the structure.

All three structures are seismic design category F and risk category III and have no structural irregularities, so the only unknown factor that could determine which analytical procedures are permitted for each structure are the fundamental period and the fundamental period over a short period.

From ASCE/SEI7 Sec. 11.4.5, because S_{D1} and S_{DS} are the same for all three structures, the fundamental period over a short period is the same for all three

structures. The fundamental period over a short period is

$$T_s = \frac{S_{D1}}{S_{DS}} = \frac{0.6}{0.75}$$
$$= 0.8\text{ sec}$$

Per ASCE/SEI7 Table 12.6-1, for a structure of seismic design category F and risk category III to qualify for the equivalent lateral-force procedure, the fundamental period must be less than

$$3.5\,T_s = (3.5)(0.8\text{ sec}) = 2.8\text{ sec}$$

The approximate fundamental period of the structures can be found using ASCE/SEI7 Eq. 12.8-7. From ASCE/SEI7 Table 12.8-2, the coefficient C_t for a steel moment-resisting frame is 0.028, and the coefficient x for a steel moment-resisting frame is 0.8. The approximate fundamental period of structure I is

$$T_a = C_t h_n^x = (0.028)(60\text{ ft})^{0.8}$$
$$= 0.74\text{ sec}$$

0.74 sec < 2.8 sec, so structure I can be analyzed using the equivalent lateral-force procedure.

The approximate fundamental period of structure II is

$$T_a = C_t h_n^x = (0.028)(160\text{ ft})^{0.8}$$
$$= 1.6\text{ sec}$$

1.6 sec < 2.8 sec, so structure II can be analyzed using the equivalent lateral-force procedure.

The approximate fundamental period of structure III is

$$T_a = C_t h_n^x = (0.028)(250\text{ ft})^{0.8}$$
$$= 2.32\text{ sec}$$

2.32 sec < 2.8 sec, so structure III can be analyzed using the equivalent lateral-force procedure.

The answer is (D).

10. Seismic inertial forces originate in all elements of a structure and are transferred through connections to horizontal diaphragms. The diaphragms distribute these forces to vertical elements of the structure (e.g., shear walls and frames) that transfer the forces into the foundation, which conveys the forces to the ground.

The answer is (D).

11. Wood structural panel diaphragms are considered flexible diaphragms that are anchored to the vertical resisting elements. The shear force in the diaphragm is distributed to the walls that are parallel to the direction of the earthquake force.

Additional shear capacity can be developed by adding nailing or stapling, overlaying the existing diaphragm with new wood structural panels, and/or providing additional shear walls or vertical bracing in the interior of a building.

The answer is (C).

12. Based on the Professional Engineers Act, Business and Professions Code, any document that does not convey an engineering decision or recommendation should not bear the seal and signature of a registered civil engineer. Reports on environmental issues, such as noise, air, hazardous waste, and so on, are necessary for the preparation of documents such as Environmental Impact Reports (EIRs) and Environmental Impact Statements (EISs). Project background and information sheets as well as drafts or incomplete documents contain factual data only and do not offer any decisions or recommendations. Therefore, option D is correct. However, if any document contains a decision or a recommendation that controls the detailed design of the project, then a seal and signature of a registered civil engineer are needed.

The answer is (D).

13. *SI Solution*

The chord force is calculated as the bending moment per unit depth of diaphragm of a simple beam under a distributed load. The maximum diaphragm chord force in the west wall occurs at midspan.

$$
C = \frac{M}{b} = \frac{wL^2}{8b} = \frac{\left(1500 \ \dfrac{\text{N}}{\text{m}}\right)(12 \ \text{m})^2}{(8)(12 \ \text{m})}
$$
$$
= 2250 \ \text{N}
$$

The shear load at each parallel wall for an earthquake force perpendicular to the south wall is

$$
V = \frac{wL}{2} = \frac{\left(3000 \ \dfrac{\text{N}}{\text{m}}\right)(12 \ \text{m})}{2}
$$
$$
= 18\,000 \ \text{N}
$$

The shear and bending moment diagrams across the length of the chord are as shown.

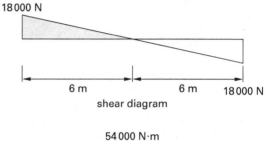

shear diagram

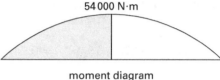

moment diagram

The moment at line 1 is calculated as the area under the shear diagram.

$$
M = (18\,000 \ \text{N})(6 \ \text{m})\left(\frac{1}{2}\right) = 54\,000 \ \text{N·m}
$$
$$
C = \frac{M}{b} = \frac{54\,000 \ \text{N·m}}{12 \ \text{m}}
$$
$$
= 4500 \ \text{N}
$$

Therefore, the chord force in the south wall is twice the chord force in the west wall.

The answer is (B).

Customary U.S Solution

The chord force is calculated as the bending moment per unit depth of diaphragm of a simple beam under a distributed load. The maximum diaphragm chord force in the west wall occurs at midspan.

$$
C = \frac{M}{b} = \frac{wL^2}{8b} = \frac{\left(100 \ \dfrac{\text{lbf}}{\text{ft}}\right)(40 \ \text{ft})^2}{(8)(40 \ \text{ft})}
$$
$$
= 500 \ \text{lbf}
$$

The shear load at each parallel wall for an earthquake force perpendicular to the south wall is

$$
V = \frac{wL}{2} = \frac{\left(200 \ \dfrac{\text{lbf}}{\text{ft}}\right)(40 \ \text{ft})}{2}
$$
$$
= 4000 \ \text{lbf}
$$

The shear and bending moment diagrams across the length of the chord are as shown.

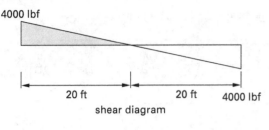

shear diagram

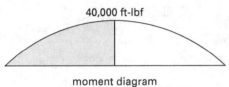

moment diagram

The moment at line 1 is calculated as the area under the shear diagram.

$$M = (4000 \text{ lbf})(20 \text{ ft})\left(\frac{1}{2}\right) = 40,000 \text{ ft-lbf}$$

$$
\begin{aligned}
C &= \frac{M}{b} \\
&= \frac{40,000 \text{ ft-lbf}}{40 \text{ ft}} \\
&= 1000 \text{ lbf}
\end{aligned}
$$

Therefore, the chord force in the south wall is twice the chord force in the west wall.

The answer is (B).

14. The largest earthquake ground motion a building is expected to experience in its life is known as the *maximum considered earthquake* and is covered in ASCE/SEI7 Sec. 11.4 and Chap. 22.

The answer is (B).

15. For a single-degree-of-freedom system, the natural period, T, is the time in which the system completes one cycle of oscillation. The natural period is the inverse of the natural frequency, f, and it can be expressed as $T = 1/f$.

For system I, the frequency is given as 0.5 Hz. The natural period for this system is

$$T_\text{I} = \frac{1}{f} = \frac{1}{0.5 \text{ Hz}} = 2 \text{ sec}$$

For system II, the natural period is 2 sec. The natural frequency for this system is

$$f_\text{II} = \frac{1}{T} = \frac{1}{2 \text{ sec}} = 0.5 \text{ Hz}$$

Therefore,

$$T_\text{I} = T_\text{II} = 2 \text{ sec}$$
$$f_\text{I} = f_\text{II} = 0.5 \text{ Hz}$$

Systems I and II both have equal natural frequencies and periods.

The answer is (B).

16. ASCE/SEI7 Sec. 12.8.2 states that the value of the structure's fundamental period, T, can be determined using the structural properties and deformational characteristics of the resisting elements in a properly substantiated analysis. This period cannot exceed the product of the coefficient for the upper limit on the calculated period, C_u, and the approximate fundamental period, T_a.

ASCE/SEI7 Sec. 12.8.2.1 provides two methods of calculating the approximate fundamental period. Equation 12.8-7 is the more commonly used equation, but Eq. 12.8-8 can be used to find the approximate fundamental period for structures that do not exceed 12 stories, that are made entirely of either steel or concrete moment-resisting frames and that have a story height of at least 10 ft (3 m). In this problem, the number of stories, N, is eight, so either equation can apply. Calculate the approximate fundamental period using both equations, and use the greater value to find the maximum fundamental period for the building.

SI Solution

For this building, the coefficient C_t is 0.0724, and the coefficient x is 0.8. Using Eq. 12.8-7, the approximate fundamental period of the building is

$$
\begin{aligned}
T_a &= C_t h_n^x = (0.0724)(36 \text{ m})^{0.8} \\
&= 1.27 \text{ s}
\end{aligned}
$$

Using Eq. 12.8-8, the approximate fundamental period of the building is

$$
\begin{aligned}
T_a &= 0.1N = (0.1)(8) \\
&= 0.8 \text{ s}
\end{aligned}
$$

Use 1.27 s as the approximate fundamental period of the building. From ASCE/SEI7 Table 12.8-1, the value of C_u for a structure with an S_{D1} value of 0.2 is 1.5. The maximum fundamental period of the building is

$$
\begin{aligned}
T &= C_u T_a = (1.5)(1.27 \text{ s}) \\
&= 1.905 \text{ s} \quad (1.9 \text{ s})
\end{aligned}
$$

The answer is (D).

Customary U.S. Solution

For this building, the coefficient C_t is 0.028, and the coefficient x is 0.8. Using Eq. 12.8-7, the approximate fundamental period of the building is

$$T_a = C_t h_n^x = (0.028)(120 \text{ ft})^{0.8}$$
$$= 1.29 \text{ sec}$$

Using Eq. 12.8-8, the approximate fundamental period of the building is

$$T_a = 0.1N = (0.1)(8)$$
$$= 0.8 \text{ sec}$$

Use 1.29 sec as the approximate fundamental period of the building. From ASCE/SEI7 Table 12.8-1, the value of C_u for a structure with an S_{D1} value of 0.2 is 1.5. The maximum fundamental period of the building is

$$T = C_u T_a = (1.5)(1.29 \text{ sec})$$
$$= 1.935 \text{ sec} \quad (1.9 \text{ sec})$$

The answer is (D).

17. The foundation sill plate is a horizontal piece of wood (redwood or treated Douglas fir) that rests on the foundation and transfers the weight of the building to the foundation. The cripple wall is a short stud wall, not a full story in height, extending from the foundation sill plate to the first floor. In the wood-frame house shown, cripple walls support the first floor of the wood structure. The exterior face is finished with siding (e.g., wood, metal, or plaster) while the studs on the inside remain exposed.

If the cripple walls are not adequately braced and strengthened, they can collapse in the event of an earthquake and the structure will fail, causing damage to the foundation, floors, walls, utility connections, and the contents of the structure. Damage may also result in fire from broken gas lines. In California, numerous cripple wall failures have been observed in previous earthquakes.

The deficient cripple walls can be strengthened in the following ways to enhance the seismic resistance of this building.

Installing expansion sill anchor bolts at regular intervals to anchor the sill plates to the foundation.

Installing steel hold-downs to anchor the wood stud walls to the foundation.

Nailing wood structural panels to the inside of the cripple studs. The top edge of the wood structural panels should be nailed into the floor framing, and the bottom edge should be nailed into the sill plate.

It should be noted that horizontal or vertical exterior siding is not strong enough to brace cripple walls.

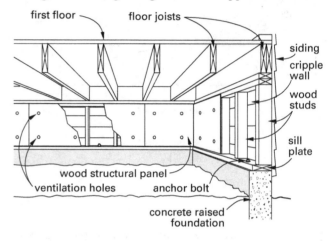

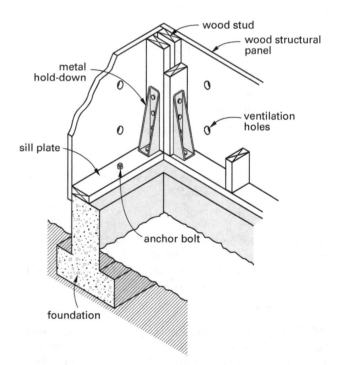

The answer is (D).

18. ASCE/SEI7 Table 11.4-1 gives $F_a = 1.2$ for $S_S = 0.75$ and site class C. ASCE/SEI7 Table 11.4-2 gives $F_v = 1.5$ for $S_1 = 0.4$ and site class C.

$$S_{MS} = F_a S_S = (1.2)(0.75) = 0.9$$
$$S_{M1} = F_v S_1 = (1.5)(0.4) = 0.6$$
$$S_{DS} = \tfrac{2}{3} S_{MS} = \left(\frac{2}{3}\right)(0.9) = 0.6$$
$$S_{D1} = \tfrac{2}{3} S_{M1} = \left(\frac{2}{3}\right)(0.6) = 0.4$$

For structure I

Use ASCE/SEI7 Eq. 12.8-1 to calculate the base shear.

$$V = C_s W$$

From ASCE/SEI7 Table 1.5-2, for risk category III, the importance factor, I_e, is 1.25. From ASCE/SEI7 Eq. 12.8-2,

$$C_s = \frac{S_{DS}}{\dfrac{R}{I_e}} = \frac{0.6}{\dfrac{8.0}{1.25}} = 0.094$$

From ASCE/SEI7 Eq. 12.8-3,

$$C_{s,\max} = \frac{S_{D1}}{T\left(\dfrac{R}{I_e}\right)} = \frac{0.40}{(0.88 \text{ sec})\left(\dfrac{8.0}{1.25}\right)} = 0.071$$

From ASCE/SEI7 Eq. 12.8-5,

$$\begin{aligned} C_{s,\min} &= 0.044 S_{DS} I_e \geq 0.01 \\ &= (0.044)(0.6)(1.25) \\ &= 0.033 \quad [> 0.01, \text{OK}] \end{aligned}$$

$C_{s,\min} = 0.033$, which is smaller than 0.071. Therefore, $C_{s,\text{gov}} = 0.071$.

For structure II

Use ASCE/SEI7 Eq. 12.8-1 to calculate the base shear.

$$V = C_s W$$

From ASCE/SEI7 Table 1.5-2, for risk category II, the importance factor, I_e, is 1.0. From ASCE/SEI7 Eq. 12.8-2,

$$\begin{aligned} C_s &= \frac{S_{DS}}{\dfrac{R}{I_e}} = \frac{0.6}{\dfrac{6.0}{1.0}} \\ &= 0.10 \end{aligned}$$

From ASCE/SEI7 Eq. 12.8-3,

$$\begin{aligned} C_{s,\max} &= \frac{S_{D1}}{T\left(\dfrac{R}{I_e}\right)} = \frac{0.40}{(0.88 \text{ sec})\left(\dfrac{6.0}{1.0}\right)} \\ &= 0.076 \end{aligned}$$

From ASCE/SEI7 Eq. 12.8-5,

$$\begin{aligned} C_{s,\min} &= 0.044 S_{DS} I_e \geq 0.01 \\ &= (0.044)(0.6)(1.0) \\ &= 0.026 \quad [> 0.01, \text{OK}] \end{aligned}$$

$C_{s,\min} = 0.026$, which is smaller than 0.076. Therefore, $C_{s,\text{gov}} = 0.076$.

Structure I has a smaller base shear than structure II.

The answer is (A).

19. IBC Sec. 202 defines a diaphragm as a horizontal or sloped system acting to transfer lateral forces to the vertical-resisting elements. Statement II and statement III are both correct.

IBC Sec. 202 explicitly states that diaphragms are systems, not subsystems. Statement I is incorrect.

The answer is (C).

20. Diaphragms, both flexible and rigid, distribute lateral forces to vertical resisting elements (e.g., columns and shear walls). Flexible diaphragms are typically of wood or light steel construction and distribute lateral forces to vertical resisting elements in proportion to the tributary area of the elements. They are incapable of distributing torsional moments to vertical resisting elements. Rigid diaphragms are typically concrete slabs or concrete metal deck floor systems. They distribute lateral forces in proportion to the rigidities of vertical resisting elements and transmit torsion to the vertical resisting elements.

The answer is (C).

21. As defined in the Professional Engineers Act, which is contained in Business and Professions Code Chap. 7, Sec. 6703 and further clarified in Rules of the Board for Professional Engineers and Professional Land Surveyors Sec. 404.1, the term "responsible charge of work" for professional engineers means the independent control and direction of the investigation or design of professional engineering work by the use of initiative, skill, decision, and judgment. However, this phrase does not refer to the concept of financial liability.

The answer is (C).

22. ASCE/SEI7 Sec. 11.2 defines a *wall system, bearing* as a structural system with bearing walls providing support for all or major portions of the vertical loads. Shear walls or braced frames provide seismic force resistance. Structure I is constructed entirely of shear walls and, therefore, lacks a complete vertical load-carrying space frame.

Structure II is a moment-resisting frame. A *moment frame* is a frame in which members and joints resist lateral forces by flexure as well as along the axis of the members. Moment frames are categorized as intermediate moment frames (IMF), ordinary moment frames (OMF), and special moment frames (SMF).

Structure III is a dual system of moment-resisting frames and shear walls. A *dual system* is a structural system with an essentially complete space frame providing support for vertical loads. Seismic force resistance is provided by moment-resisting frames and shear walls or braced frames.

The answer is (A).

23. The Field Act requires special seismic design for public school buildings, subject to the approval of the Division of the State Architect, Department of General Services. Older school buildings need to be reevaluated and reassessed in light of modern building codes. The provisions of the Field Act do not apply to private schools and colleges. The act that mandates hospitals to be fully operational and functional after an earthquake is the California Hospital Act.

The answer is (A).

24. ρ represents a reliability/redundancy factor that should be assigned to all structures according to ASCE/SEI7 Sec. 12.3.4. This factor is based on the extent of structural redundancy inherent in the design configuration of the structure and its lateral force-resisting systems.

SI Solution

x is the length of the south wall. For an east-west earthquake, the diaphragm shear is carried equally between the north and south walls.

$$V = \frac{wL}{2} = \frac{\left(4400 \ \dfrac{N}{m}\right)(12 \ m)}{2}$$
$$= 26\,400 \ N$$

The diaphragm shear stress is the diaphragm shear divided by the depth of the diaphragm.

$$v = \frac{V}{b}$$
$$v_{roof} = \frac{26\,400 \ N}{15 \ m}$$
$$= 1760 \ N/m$$
$$v_{wall} = \frac{26\,400 \ N}{x}$$

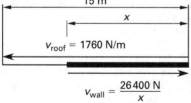

The drag strut carries the diaphragm load for the area tributary to the strut.

$$D_x + v_{roof}x - v_{wall}x = 0$$
$$8900 \ N + \left(1760 \ \frac{N}{m}\right)x - \left(\frac{26\,400 \ N}{x}\right)x = 0$$
$$x \approx 10 \ m$$

The answer is (B).

Customary U.S. Solution

x is the length of the south wall. For an east-west earthquake, the diaphragm shear is carried equally between the north and south walls.

$$V = \frac{wL}{2}$$
$$= \frac{\left(300 \ \dfrac{lbf}{ft}\right)(40 \ ft)}{2}$$
$$= 6000 \ lbf$$

The diaphragm shear stress is the diaphragm shear divided by the depth of the diaphragm.

$$v = \frac{V}{b}$$
$$v_{roof} = \frac{6000 \ lbf}{50 \ ft}$$
$$= 120 \ lbf/ft$$
$$v_{wall} = \frac{6000 \ lbf}{x}$$

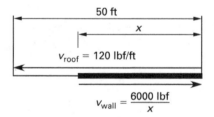

The drag strut carries the diaphragm load for the area tributary to the strut.

$$D_x + v_{roof}x - v_{wall}x = 0$$
$$2000 \ lbf + \left(120 \ \frac{lbf}{ft}\right)x - \left(\frac{6000 \ lbf}{x}\right)x = 0$$
$$x \approx 30 \ ft$$

The answer is (B).

$$\frac{6000}{50} \ -$$

25. *SI Solution*

The overturning moment, OTM, is given by

$$\text{OTM} = Vh$$

The base shear is

$$V = mS_a$$

The response spectra indicate that the maximum S_a/g occurs between $T = 0.8$ s and $T = 1.2$ s.

For $T = 0.8$ s, h is 15 m and $S_a/g = 0.6$.

$$V = mS_a$$

$$= \frac{(9000 \text{ kg})\left(9.81 \ \dfrac{\text{m}}{\text{s}^2}\right)(0.6)}{1000 \ \dfrac{\text{N}}{\text{kN}}}$$

$$= 52.97 \text{ kN}$$

$$\text{OTM} = Vh = (52.97 \text{ kN})(15 \text{ m})$$

$$= 794.6 \text{ kN·m}$$

For $T = 1.2$ s, h is 23 m and $S_a/g = 0.6$.

$$V = mS_a$$

$$= \frac{(9000 \text{ kg})\left(9.81 \ \dfrac{\text{m}}{\text{s}^2}\right)(0.6)}{1000 \ \dfrac{\text{N}}{\text{kN}}}$$

$$= 52.97 \text{ kN}$$

$$\text{OTM} = Vh = (52.97 \text{ kN})(23 \text{ m})$$

$$= 1218 \text{ kN·m} \quad (1200 \text{ kN·m})$$

W	h (m)	T (s)	S_a/g	$V = mS_a$ (kN)	$\text{OTM} = Vh$ (kN·m)
9000 kg	30	2.0	0.40	35.32	1060
	23	1.2	0.60	52.97	1218
	15	0.8	0.60	52.97	795
	7.6	0.3	0.45	39.73	302

The maximum overturning moment is 1218 kN·m (1200 kN·m).

The answer is (C).

Customary U.S. Solution

The overturning moment, OTM, is given by

$$\text{OTM} = Vh$$

The base shear is

$$V = mS_a$$

The response spectra indicate that the maximum S_a/g occurs between $T = 0.8$ sec and $T = 1.2$ sec.

For $T = 0.8$ sec, h is 50 ft and $S_a/g = 0.6$.

$$V = mS_a = W\left(\frac{S_a}{g}\right) = (20 \text{ kips})(0.6)$$

$$= 12 \text{ kips}$$

$$\text{OTM} = Vh = (12 \text{ kips})(50 \text{ ft})$$

$$= 600 \text{ ft-kips}$$

For $T = 1.2$ sec, h is 75 ft and $S_a/g = 0.6$.

$$V = mS_a = W\left(\frac{S_a}{g}\right)$$

$$= (20 \text{ kips})(0.6)$$

$$= 12 \text{ kips}$$

$$\text{OTM} = Vh = (12 \text{ kips})(75 \text{ ft})$$

$$= 900 \text{ ft-kips}$$

W	h (ft)	T (sec)	S_a/g	$V = mS_a$ (kips)	$\text{OTM} = Vh$ (ft-kips)
20 kips	100	2.0	0.40	8	800
	75	1.2	0.60	12	900
	50	0.8	0.60	12	600
	25	0.3	0.45	9	225

The maximum overturning moment is 900 ft-kips.

The answer is (C).

26. The most effective technique for decreasing the acceleration imposed on the building system is to install seismic base isolators at the base of the building. In an earthquake, the ground moves independently of buildings that are base isolated. Therefore, the acceleration decreases while the fundamental period of the building system increases. As a result, the base shear of the structural system of the building will be reduced. Regulations for seismic-isolated structures are provided in ASCE/SEI7 Chap. 17.

Base isolators function effectively on soil profiles with bedrock or firm soils. They are not suitable for soil profile type S_E, where the soil is soft. The long period of soft soil may coincide with the increased period of the structure, resulting in a resonant condition.

In addition to seismic base isolation, other energy dissipation devices may be added to the structural system, primarily to increase the system damping.

Improving the energy-absorbing capacity of a structure by modifying, removing, or replacing its existing structural system is seldom economically feasible. There are a few exceptions: An ordinary steel moment frame can be upgraded to a special moment frame, and concentric steel bracing or unreinforced masonry infill walls can be modified by removing the bracing or the infill walls and installing eccentric bracing or reinforced concrete shear walls.

The answer is (A).

27. *SI Solution*

$$b = 4.6 \text{ m} + 7.6 \text{ m}$$
$$= 12.2 \text{ m}$$
$$V = \left(7300 \ \frac{\text{N}}{\text{m}}\right)(12.2 \text{ m})$$
$$= 89\,060 \text{ N}$$

Alternatively,

$$V_{\text{panel I}} = \left(7300 \ \frac{\text{N}}{\text{m}}\right)(4.6 \text{ m})$$
$$= 33\,580 \text{ N}$$
$$V_{\text{panel II}} = \left(7300 \ \frac{\text{N}}{\text{m}}\right)(7.6 \text{ m})$$
$$= 55\,480 \text{ N}$$

$$V = V_{\text{panel I}} + V_{\text{panel II}}$$
$$= 33\,580 \text{ N} + 55\,480 \text{ N}$$
$$= 89\,060 \text{ N} \quad (89\,000 \text{ N})$$

The answer is (B).

Customary U.S. Solution

$$b = 15 \text{ ft} + 25 \text{ ft}$$
$$= 40 \text{ ft}$$
$$V = \left(500 \ \frac{\text{lbf}}{\text{ft}}\right)(40 \text{ ft})$$
$$= 20{,}000 \text{ lbf}$$

Alternatively,

$$V_{\text{panel I}} = \left(500 \ \frac{\text{lbf}}{\text{ft}}\right)(15 \text{ ft})$$
$$= 7500 \text{ lbf}$$
$$V_{\text{panel II}} = \left(500 \ \frac{\text{lbf}}{\text{ft}}\right)(25 \text{ ft})$$
$$= 12{,}500 \text{ lbf}$$
$$V = V_{\text{panel I}} + V_{\text{panel II}}$$
$$= 7500 \text{ lbf} + 12{,}500 \text{ lbf}$$
$$= 20{,}000 \text{ lbf}$$

The answer is (B).

28. The provision "Practice within Area of Competence," Sec. 415 of the Rules of the Board, specifies that a professional registered engineer can practice and perform engineering work only in the field(s) in which the engineer is educated and/or experienced and is fully competent and proficient.

The Rules prohibit a professional highway engineer from being in responsible charge in areas other than those in which the engineer is qualified. Therefore, the engineer cannot design high-rises, hospitals, or highway bridge structures unless the engineer is fully competent in those areas. (Competency and proficiency in highway and transportation design do not qualify a civil PE to be in responsible charge of the design of a highway bridge structure.) The responsible charge for designing hospitals in California is restricted to licensed structural engineers [Health and Safety Code, Sec. 129805(a)]. In some areas of California, the design of high-rises is also restricted to licensed structural engineers. Where these restrictions apply, a professional highway engineer can perform engineering design work only under the direction of the licensed structural engineer who is responsible for signing final plans.

The answer is (B).

29. The NEHRP publication, *Recommended Provisions for Seismic Regulations for New Buildings and Other Structures*, states that the primary goal of seismic design is

1. to provide minimum design criteria for structures appropriate to their primary function and use considering the need to protect the health, safety, and welfare of the general public by minimizing the earthquake-related risk to life

2. to improve the capability of essential facilities and structures containing substantial quantities of hazardous materials to function during and after design earthquakes

Damage control (i.e., preventing structure failures or preserving property) is not the primary focus of seismic design.

The answer is (A).

30. Ductility is the ability of a material to yield without collapse. The structure response modification factor, R, is a measure of inherent overstrength and global ductility and is determined from the type of structural system selected. It is defined for buildings in ASCE/SEI7 Table 12.2-1 and for nonbuilding structures in ASCE/SEI7 Table 15.4-1 and Table 15.4-2.

When a combination of different structural systems is used to resist lateral forces in the same direction, the value of R per ASCE/SEI7 Sec. 12.2.3.2 is used. When a combination of different structural systems is used vertically, the value of R per ASCE/SEI7 Sec. 12.2.3.1 is used. ASCE/SEI7 Sec. 12.2.4 has additional information.

The answer is (D).

31. To determine the minimum value of the base shear, the minimum value of the seismic coefficient must be determined. From ASCE/SEI7 Eq. 12.8-5, buildings must have a minimum coefficient of at least 0.01. However, when S_1 is greater than 0.6, ASCE/SEI7 Eq. 12.8-6 must also be considered.

$$V = \left(\frac{0.5 S_1}{\dfrac{R}{I_e}} \right) W$$

The answer is (B).

32. The eccentricity is the distance between the centers of mass and rigidity measured in a direction perpendicular to the lateral force. According to ASCE/SEI7 Sec. 12.8.4.2, an accidental eccentricity, e_a, defined as 5% of the building dimension perpendicular to the lateral force, should be assumed for determining the horizontal distribution of story shear in structures with rigid diaphragms.

For the N-S loading,

$$e_a = \pm 0.05 L$$

For the E-W loading,

$$e_a = \pm 0.05 d$$

Therefore, cases I and IV appropriately account for accidental eccentricity in the north-south and east-west loading directions, respectively.

The answer is (C).

33. Liquefaction is the result of an increase in pore water pressure and a decrease in the effective stress of a soil, and it can lead to a complete loss of bearing capacity. A site with predominantly loose, saturated, cohesionless soil (i.e., sands and silts) is most likely to experience liquefaction in the event of a strong earthquake.

A soil profile containing a deep clay layer, site class E, will not undergo liquefaction, but may be susceptible to ground motion amplification in the event of an earthquake. Isolated spread footings supported on a liquefiable soil profile may experience differential vertical and lateral settlements, loss of bearing capacity, and overturning forces. The structure may be damaged, and utility connections to the building may be disrupted. Lateral spreading is also a concern if the liquefiable soils are located on a gradient, however slight.

Structures on foundations of piles or drilled piers that penetrate through liquefaction-susceptible soils and bear on site class A or B (i.e., bedrock) can be designed for adequate performance during an earthquake. The foundation should be rigidly tied to the structure and should be capable of transferring the seismic shear from the pile caps through the piles to the bearing soil when lateral support to the piles is significantly reduced due to liquefaction. To increase the level of protection and safety, in addition to an adequate foundation, the design lateral force, the redundancy, the value of R, and the quality of the construction materials and construction methods should be increased. Where preliminary analysis indicates a foundation problem, a qualified geotechnical engineer should establish criteria for the foundation analysis. ASCE/SEI7 Sec. 11.8.3, item 2 provides additional recommendations to address the potential for liquefaction.

The answer is (A).

34. The structure uses one structural system to resist forces in the x-direction; it uses two structural systems to resist forces in the y-direction. Dual systems are defined in ASCE/SEI7 Sec. 11.2 and refer to two different structural systems in the same x- or y-direction at the same story level. Neither the concept or nor the term "triple system" is defined in ASCE/SEI7.

According to ASCE/SEI7 Table 12.2-1, ordinary reinforced concrete shear wall bearing systems are not permitted (NP) in seismic design category D.

ASCE/SEI7 Sec. 12.8.4.1 requires all buildings with rigid diaphragms to consider accidental torsion.

The answer is (C).

35. *SI Solution*

For column B, if one end is pinned, simple cantilever curvature occurs. The stiffness for this center column is

$$
\begin{aligned}
K_B &= \frac{3EI}{L^3} \\
&= \frac{(3)(2.0 \times 10^5 \text{ MPa})(8.1 \times 10^{-6} \text{ m}^4)\left(10^6 \, \dfrac{\text{Pa}}{\text{MPa}}\right)}{(6.1 \text{ m})^3} \\
&= 2.14 \times 10^4 \text{ N/m}
\end{aligned}
$$

(1 Pa is equivalent to 1 N/m^2.)

Columns A and C are rigidly fixed at both ends. Therefore, the stiffness for these columns is

$$K_A = K_C$$

$$K = \frac{12EI}{L^3}$$

$$= \frac{(12)(2.0 \times 10^5 \text{ MPa})(8.1 \times 10^{-6} \text{ m}^4)\left(10^6 \frac{\text{Pa}}{\text{MPa}}\right)}{(6.1 \text{ m})^3}$$

$$= 8.56 \times 10^4 \text{ N/m}$$

$$K_{\text{total}} = K_A + K_B + K_C$$

$$= \left(8.56 \times 10^4 \frac{\text{N}}{\text{m}}\right) + \left(2.14 \times 10^4 \frac{\text{N}}{\text{m}}\right)$$

$$+ \left(8.56 \times 10^4 \frac{\text{N}}{\text{m}}\right)$$

$$= 19.26 \times 10^4 \text{ N/m}$$

The resisting force in column B is proportional to its relative stiffness.

$$F_B = \left(\frac{K_B}{K_{\text{total}}}\right)(490 \text{ kN})$$

$$= \left(\frac{2.14 \times 10^4 \frac{\text{N}}{\text{m}}}{19.26 \times 10^4 \frac{\text{N}}{\text{m}}}\right)(490 \text{ kN})$$

$$= 54.44 \text{ kN} \quad (50 \text{ kN})$$

The answer is (A).

Customary U.S. Solution

For column B, if one end is pinned, simple cantilever curvature occurs. The stiffness for this center column is

$$K_B = \frac{3EI}{L^3}$$

$$= \frac{(3)\left(29 \times 10^6 \frac{\text{lbf}}{\text{in}^2}\right)(19.5 \text{ in}^4)}{(20 \text{ ft})^3 \left(12 \frac{\text{in}}{\text{ft}}\right)^3}$$

$$= 122.72 \text{ lbf/in}$$

Columns A and C are rigidly fixed at both ends. Therefore, the stiffness for these columns is

$$K_A = K_C$$

$$K = \frac{12EI}{L^3}$$

$$= \frac{(12)\left(29 \times 10^6 \frac{\text{lbf}}{\text{in}^2}\right)(19.5 \text{ in}^4)}{(20 \text{ ft})^3 \left(12 \frac{\text{in}}{\text{ft}}\right)^3}$$

$$= 490.89 \text{ lbf/in}$$

$$K_{\text{total}} = K_A + K_B + K_C$$

$$= 490.89 \frac{\text{lbf}}{\text{in}} + 122.72 \frac{\text{lbf}}{\text{in}} + 490.89 \frac{\text{lbf}}{\text{in}}$$

$$= 1104.5 \text{ lbf/in}$$

The resisting force in column B is proportional to its relative stiffness.

$$F_B = \left(\frac{K_B}{K_{\text{total}}}\right)(110 \text{ kips})$$

$$= \left(\frac{122.72 \frac{\text{lbf}}{\text{in}}}{1104.5 \frac{\text{lbf}}{\text{in}}}\right)(110 \text{ kips})$$

$$= 12.22 \text{ kips} \quad (10 \text{ kips})$$

The answer is (A).

36. Based on Chap. 7, Sec. 6737.1 of the Professional Engineers Act, California Business and Professions Code, "Professional Engineers," the preparation of plans, drawings, or specifications for (1) single-family dwellings of wood-frame construction not more than two stories and a basement in height and (2) multiple dwellings consisting of no more than four dwelling units (e.g., apartment or condominium complexes) of wood-frame construction not more than two stories and a basement in height can be performed by any person, including qualified architects, civil engineers, and structural engineers.

The answer is (D).

37. IBC Table 1613.2.3(1) and Table 1613.2.3(2) provide the values for the site coefficients F_a and F_v. Straight-line interpolation must be used for intermediate values.

For $S_S = 0.75$, $F_a = 1.2$.

For $S_S = 1.00$, $F_a = 1.1$. Therefore, for $S_S = 0.9$,

$$F_a = 1.2 - \left(\frac{0.9 - 0.75}{1.00 - 0.75} \right)(1.1 - 1.2) = 1.14$$

For S_1 0.5, $F_v = 1.8$.

The answer is (D).

38. P-delta effects are related to the magnitude of the additional overturning moment that is generated when the drifts are large. Δ is the design story drift corresponding to the story shear, V_x. ASCE/SEI7 Sec. 12.8.7 provides the conditions when P-delta effects need to be considered. They do not need to be considered when the stability coefficient, θ, is less than or equal to 0.10.

The answer is (A).

39. *SI Solution*

For the east shear wall, the lateral force due to shear is

$$F_v = V \left(\frac{R_{\text{tab,east}}}{\sum R_{\text{tab},i}} \right)$$
$$= (220\,000 \text{ N}) \left(\frac{3}{3 + 3} \right)$$
$$= 110\,000 \text{ N}$$

The moment due to torsion is

$$T_{\text{N-S}} = Ve \qquad \text{[Eq. 1]}$$

The actual eccentricity is

$$e = \text{CM} - \text{CR} = 13.7 \text{ m} - 9.1 \text{ m}$$
$$= 4.6 \text{ m}$$

The accidental eccentricity required by ASCE/SEI7 Sec. 12.8.4.2 is 5% of the building dimension perpendicular to the lateral force.

$$e_a = (0.05)(27.4 \text{ m})$$
$$= 1.37 \text{ m}$$

The total eccentricity is

$$e_{\text{N-S}} = e + e_a$$
$$= 4.6 \text{ m} + 1.37 \text{ m}$$
$$= 5.97 \text{ m}$$

Using Eq. 1, the moment due to torsion is

$$T_{\text{N-S}} = Ve_{\text{N-S}} = \frac{(220\,000 \text{ N})(5.97 \text{ m})}{1000 \, \frac{\text{N}}{\text{kN}}}$$
$$= 1313 \text{ kN·m}$$

The lateral force due to torsion is

$$F_t = T_{\text{N-S}} \left(\frac{R_{\text{tab,east}} d}{\sum R_{\text{tab},i} d^2} \right) \qquad \text{[Eq. 2]}$$

The sum of $R_{\text{tab},i} d^2$ can be obtained from the table.

$$\sum R_{\text{tab},i} d^2 = 225 \text{ m}^2 + 127 \text{ m}^2 + 1000 \text{ m}^2$$
$$+ 248 \text{ m}^2$$
$$= 1600 \text{ m}^2$$

Using Eq. 2, the lateral force due to torsion is

$$F_t = T_{\text{N-S}} \left(\frac{R_{\text{tab,east}} d}{\sum R_{\text{tab},i} d^2} \right)$$
$$= (1313 \text{ kN·m}) \left(\frac{54.9 \text{ m}}{1600 \text{ m}^2} \right) \left(1000 \, \frac{\text{N}}{\text{kN}} \right)$$
$$= 45\,052 \text{ N}$$

The lateral force resultant acts through the center of mass, while the resisting force resultant acts through the center of rigidity. Therefore, for an earthquake motion in the north direction, a torsional moment develops that has counterclockwise rotation; the building acts as though pinned at the center of rigidity.

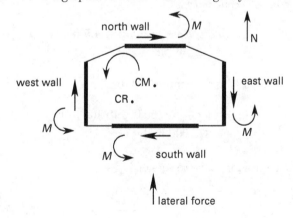

The east wall receives positive torsional components.

Therefore, the total lateral force is the sum of the shear force and the torsional force.

$$F_{\text{total}} = F_v + F_t = 110\,000 \text{ N} + 45\,052 \text{ N}$$
$$= 155\,052 \text{ N} \quad (155\,000 \text{ N})$$

The answer is (D).

Customary U.S. Solution

For the east shear wall, the lateral force due to shear is

$$F_v = V\left(\frac{R_{\text{tab,east}}}{\sum R_{\text{tab},i}}\right)$$
$$= (50{,}000 \text{ lbf})\left(\frac{3}{3+3}\right)$$
$$= 25{,}000 \text{ lbf}$$

The moment due to torsion is

$$T_{\text{N-S}} = Ve \qquad \text{[Eq. 1]}$$

The actual eccentricity is

$$e = \text{CM} - \text{CR}$$
$$= 45 \text{ ft} - 30 \text{ ft}$$
$$= 15 \text{ ft}$$

The accidental eccentricity required by ASCE/SEI7 Sec. 12.8.4.2 is 5% of the building dimension perpendicular to the lateral force.

$$e_a = (0.05)(90 \text{ ft})$$
$$= 4.5 \text{ ft}$$

The total eccentricity is

$$e_{\text{N-S}} = e + e_a = 15 \text{ ft} + 4.5 \text{ ft}$$
$$= 19.5 \text{ ft}$$

Using Eq. 1, the moment due to torsion is

$$T_{\text{N-S}} = Ve_{\text{N-S}} = (50{,}000 \text{ lbf})(19.5 \text{ ft})$$
$$= 975{,}000 \text{ ft-lbf}$$

The lateral force due to torsion is

$$F_t = T_{\text{N-S}}\left(\frac{R_{\text{tab,east}}d}{\sum R_{\text{tab},i}d^2}\right) \qquad \text{[Eq. 2]}$$

The sum of $R_{\text{tab},i}d^2$ can be obtained from the table.

$$\sum R_{\text{tab},i}d^2 = 2450 \text{ ft}^2 + 1350 \text{ ft}^2 + 10{,}800 \text{ ft}^2 + 2700 \text{ ft}^2$$
$$= 17{,}300 \text{ ft}^2$$

Using Eq. 2, the lateral force due to torsion is

$$F_t = T_{\text{N-S}}\left(\frac{R_{\text{tab,east}}d}{\sum R_{\text{tab},i}d^2}\right) = (975{,}000 \text{ ft-lbf})\left(\frac{180 \text{ ft}}{17{,}300 \text{ ft}^2}\right)$$
$$= 10{,}145 \text{ lbf}$$

The lateral force resultant acts through the center of mass, while the resisting force resultant acts through the center of rigidity. Therefore, for an earthquake motion in the north direction, a torsional moment develops that has counterclockwise rotation; the building acts as though pinned at the center of rigidity.

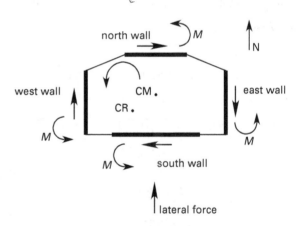

The east wall receives positive torsional components. Therefore, the total lateral force is the sum of the shear force and the torsional force.

$$F_{\text{total}} = F_v + F_t$$
$$= 25{,}000 \text{ lbf} + 10{,}145 \text{ lbf}$$
$$= 35{,}145 \text{ lbf} \quad (35{,}000 \text{ lbf})$$

The answer is (D).

40. The Professional Engineers Act, California Business and Professions Code, "Professional Engineers," Chap. 7 requires that any person who wishes to practice civil engineering must be appropriately registered. An applicant should submit an application to the Board to take the appropriate examinations for registration. The Board evaluates the applicant's professional experience and education and decides whether the applicant meets the Board's registration requirements. The applicant's experience and education should evidence that the engineer is competent to practice civil engineering based on Sec. 6750 through Sec. 6762 of the Professional Engineers Act. A civil engineering degree is not a requirement for registration qualification. However, higher education can reduce the Board's number of years of engineering experience requirements [Sec. 6753].

Registered engineers can use the titles "Professional Engineer," "Registered Engineer," "Consulting Engineer," or any combination of those titles. The provisions of the California Business and Professions Code also require that "Professional Engineers" who practice civil engineering be qualified according to the rules and regulations established for civil engineers by the Board.

The answer is (C).

41. ASCE/SEI7 Sec. 12.1.3 requires all parts of the structure between separation joints be interconnected to form a continuous path to the lateral force-resisting systems. The connection strength must be at least 5% of the weight of the smaller portion.

The answer is (B).

42. *SI Solution*

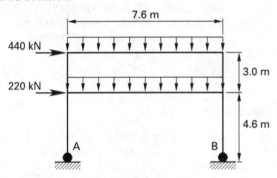

Determine the moment about point B due to the applied forces. Consider clockwise moments to be positive.

$$\sum M_B = 0$$

$$
(440 \text{ kN})(7.6 \text{ m})
$$
$$
+ (220 \text{ kN})(4.6 \text{ m})
$$
$$
- (2)\left(\frac{1}{2}\right)\left(29 \ \frac{\text{kN}}{\text{m}}\right)(7.6 \text{ m})^2
$$
$$
- F_A(7.6 \text{ m}) = 0
$$
$$
F_A = 352.8 \text{ kN} \quad (350 \text{ kN})
$$

The answer is (B).

Customary U.S. Solution

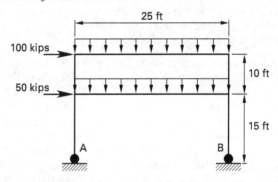

Determine the moment about point B due to the applied forces. Consider clockwise moments to be positive.

$$\sum M_B = 0$$

$$
(100 \text{ kips})(25 \text{ ft}) + (50 \text{ kips})(15 \text{ ft})
$$
$$
- (2)\left(\frac{1}{2}\right)\left(2 \ \frac{\text{kips}}{\text{ft}}\right)(25 \text{ ft})^2
$$
$$
- F_A(25 \text{ ft}) = 0
$$
$$
F_A = 80 \text{ kips}
$$

The answer is (B).

43. The shear capacity of existing steel decking is increased by additional welding to the steel ledger and by adding a reinforced concrete fill. The connection to the wall is deficient for in-plane shear and out-of-plane anchorage forces.

To increase the shear transfer capacity of this connection, an additional steel ledger can be provided to connect the steel support to the wall. The new steel ledger should be drilled, grouted, and bolted to the wall. It should also be welded to the existing steel ledger.

To provide the additional out-of-plane anchorage capacity, a new steel strap can be installed. The new steel strap should be welded to the new ledger and to the underside of the steel decking.

The answer is (C).

44. Typically, construction costs of a building can be estimated by the type of building materials selected and the square footage (m^2) of building space involved in its construction. Building materials influence the direct cost of the amount of work needed to complete the project. Indirect costs, such as permits, fees, construction financing, and so on, should not be included in this estimate. It should be noted that structural type also influences the construction cost of buildings. More ductile structures, which are reflected in high R factors, are associated with increased building construction costs. For example, construction costs for steel with $R = 8.5$ are higher than construction costs for wood-framed structures with $R = 6.5$.

Structural steel is a highly ductile material, which makes it relatively expensive. The speed of construction for structural steel is fast, depending on the amount of welding or connections required.

Concrete construction costs are generally lower than steel construction costs. R for concrete structural systems varies from 1.5 to 8.0 [ASCE/SEI7 Table 12.2-1]. Concrete construction is slow because it necessitates false (form) work and curing periods. In addition, use of ductile concrete is essential in seismically active areas, and much attention should be given to connection and confinement details. Concrete structures may be cast-in-place, precast, or a combination of the two. Concrete "tilt-up" buildings can be built relatively quickly, but tilt-up construction is mainly used in one-story industrial and commercial buildings.

The strength of masonry structures is sensitive to materials, design, construction, and quality control of masonry. Masonry construction costs are lower than those of concrete and steel. Masonry construction is relatively fast compared to concrete construction because masonry construction requires either minimal or no false work.

Well-designed wood structures in high seismic areas have generally had a good seismic performance record, and many residential structures are built of this construction type. The cost of wood construction is lower than that of other building materials. The speed of construction is very fast.

A comparison among the typical cost ranges as well as the corresponding speed of construction for the four different building materials is given in the table shown. Wood exhibits the lowest typical construction cost as well as the fastest type of construction.

construction	steel	concrete	masonry	wood
cost	expensive	relatively expensive	relatively expensive	less expensive
speed	fast	slow	relatively fast	very fast

The answer is (D).

45. There are advanced structural engineering computer programs for structural analysis and design. Analysis and design can be done on structures constructed of any material or combination of materials. These computer programs introduce a technology that provides an advanced level of design capability for frames, trusses, continuous beams, columns, shear walls, frame/shear wall combinations, and more. The analysis options they provide are complete static analysis, P-delta analysis, dynamic analysis, time history analysis, response spectra analysis, and much more.

New and enhanced computer programs continue to bring new power and capabilities to the engineering profession. Notable among the computer programs are RISA-3D, SAP2000, STAAD.Pro, Ram Structural System, CSI Perform 3D, and ETABS.

The answer is (D).

46. Chords are boundary members of a diaphragm, such as walls or reinforcements, that are perpendicular to the direction of applied lateral load. They resist moments and are regarded as tension and compression members. The maximum chord forces occur at midspan where the maximum moment develops. Chord forces at the chord ends are zero.

Struts are perimeter members of a diaphragm that are parallel to the direction of the applied load. At points of discontinuity in the plans, or where the vertical resisting elements are not provided because of an opening (i.e., windows or doors), the drag struts collect diaphragm load and transfer it to the supporting vertical resisting elements. Struts carry tension and compression, depending on the direction of applied lateral load. The maximum force develops at the point where the drag strut attaches to the parallel wall. The outside ends of the walls experience no strut forces.

The maximum chord force and strut force are never developed at the same location.

The answer is (D).

47. It takes 32 small earthquakes to radiate the same energy as an earthquake one magnitude larger. Once the Richter magnitude, M, is known, an approximate correlation (the Gutenberg-Richter law) can be used to determine the radiated energy in ergs.

$$\log_{10} E = 11.8 + 1.5M$$

Assume that $M_1 = 6.0$ and $M_2 = 5.0$. Then, substitute M_1 and M_2 values into the equation and solve.

$$\log_{10} E_1 = 11.8 + 1.5M_1 = 11.8 + (1.5)(6.0)$$
$$E_1 = 6.3096 \times 10^{20} \text{ ergs}$$
$$\log_{10} E_2 = 11.8 + 1.5M_2 = 11.8 + (1.5)(5.0)$$
$$E_2 = 1.9953 \times 10^{19} \text{ ergs}$$
$$\frac{E_1}{E_2} = \frac{6.3096 \times 10^{20} \text{ ergs}}{1.9953 \times 10^{19} \text{ ergs}} = 32$$

An earthquake of magnitude 6.0 would radiate approximately 32 times more energy than an earthquake just one magnitude smaller.

The answer is (C).

48. ASCE/SEI7 Sec. 12.2.2 allows different response modification factor, R, values for the x- and y-directions of a structure. ASCE/SEI7 Sec. 12.2.3.1 explains how to determine the R value if two different systems are used for different levels of a structure in the same direction (e.g., the y-direction). When the upper system has a lower value of R, this lower value should be used for the entire height of the structure. For the two steel systems shown, ASCE/SEI7 Table 12.2-1 lists an R value of 6 for the steel special concentrically braced frame (the upper system), and a value of 8 for the steel special moment resisting frame, (the lower system). Use the lower value of 6 for R in the y-direction.

The answer is (B).

49. Based on IBC Table 2306.3(1), a maximum allowable shear stress of 315 lbf/ft (4600 N/m) will be achieved when the staple spacing at the wood structural panel edges is 3 in (76 mm).

The answer is (B).

50. According to IBC Sec. 202 (Definitions), an approved agency is approved by the local building official or authority having jurisdiction for the purpose of conducting materials testing or inspection of civil engineering projects.

IBC Sec. 202 also states that an inspection certificate is applied on a product or material to indicate that it has been inspected and evaluated by an approved agency. An inspection certificate cannot be applied to an entire building.

According to IBC Sec. 202 (Definitions), continuous special inspection must be performed by a special inspector who is present when and where the work to be inspected is being done. A continuous special inspection may be performed by a part-time inspector as long as he or she fulfills this requirement.

IBC Sec. 202 (Definitions) explicitly does not permit structural observation to include or waive the responsibility for work performed by a qualified inspector.

The answer is (C).

51. IBC Sec. 1705.4.2 states that special inspections are required for vertical masonry foundation elements. IBC Table 1705.3 states that special inspections are required for anchors cast in concrete. IBC 1705.2.2 states that special inspections are required for cold-formed steel decks. Therefore, all three types of construction would require special inspection.

The answer is (D).

52. A diaphragm anchor resists an out-of-plane seismic load by connecting the structural wall to the roof diaphragm. The in-plane shear flow loads are not relevant. The design force on individual anchors is calculated in accordance with ASCE/SEI7 Sec. 12.11 and requires knowing the weight of the tributary portion of the wall.

The redundancy factor is not needed to calculate the anchor design force.

The answer is (A).

53. ASCE/SEI7 Table 12.3-2 defines both weak and soft stories as types of vertical irregularities. Soft-story irregularity is related to story stiffness, and is not related to story strength (i.e., load-carrying ability). According to items 5a and 5b of ASCE/SEI7 Table 12.3-2, weak stories must be checked against the lateral strength of the story directly above it. Per item 5a of ASCE/SEI7 Table 12.3-2, a weak story exists if the lateral strength of a story is greater than 66% but less than 80% of the strength of the story directly above it. Under item 5b of ASCE/SEI7 Table 12.3-2, an extreme weak story exists if the lateral strength of the story is less than 65% of the strength of the story directly above it.

SI Solution

The first level of the building is the potential soft story. This first level must be checked against the story directly above it. The 80% and 65% limit values for the lateral strength of the lower story are

$$V_{\text{weak}} = (0.80)(1100 \text{ kN}) = 880 \text{ kN}$$
$$V_{\text{ext-weak}} = (0.65)(1100 \text{ kN}) = 715 \text{ kN}$$

Since the lower story strength of 800 kN is less than the 80% limit of 880 kN, but more than the 65% limit of 715 kN, the building is defined as a weak story irregularity.

The answer is (C).

Customary U.S. Solution

The first level of the building is the potential soft story. This first level must be checked against the story directly above it. The 80% and 65% limit values for the lateral strength of the lower story are

$$V_{\text{weak}} = (0.80)(250 \text{ kips}) = 200 \text{ kips}$$
$$V_{\text{ext-weak}} = (0.65)(250 \text{ kips}) = 163 \text{ kips}$$

Since the lower story lateral strength of 180 kips is less than the 80% limit of 200 kips but more than the 65% limit of 163 kips, the building has a weak story irregularity.

The answer is (C).

54. ASCE/SEI7 Sec. 12.2 describes eight major types of structural systems (bearing wall, building frame, moment-resisting frame, dual systems with special moment-resisting frames, dual systems with intermediate moment-resisting frames, shear wall-frame, cantilevered column, and steel systems without seismic detailing). The factors used to determine a structural system's base shear are the response modification factor, R, the overstrength factor, Ω_O, and the deflection amplification factor, C_d. The redundancy factor, ρ, is not used to calculate base shear.

The answer is (D).

55. ASCE/SEI7 Sec. 12.8.2 gives two methods to determine the fundamental period of vibration of a building. The first method is an approximate method, described in ASCE/SEI7 Sec. 12.8.2.1, while the other method uses computer analysis. When the fundamental period of vibration is calculated using computer analysis, it is usually longer than the fundamental period calculated using the approximate method. ASCE/SEI7 Sec. 12.8.2 sets an upper limit on the period that may be used to determine the design base shear, no matter what the period is as determined by the computer analysis.

SI Solution

Use ASCE/SEI7 Eq. 12.8-7 to determine the approximate value for the period of vibration of the building. From ASCE/SEI7 Table 12.8-2, C_t is 0.0731 and x is 0.75 for an eccentrically based steel frame.

$$
\begin{aligned}
T_a &= C_t h_n^x \\
&= (0.0731)(50 \text{ m})^{0.75} \\
&= 1.37 \text{ sec}
\end{aligned}
$$

ASCE/SEI7 Sec. 12.8.2 defines the upper limit for the fundamental period as the product of C_u and T_a. From ASCE/SEI7 Table 12.8-1, C_u is 1.4 for buildings with S_{D1} values greater than 0.4. The upper-limit value is

$$
\begin{aligned}
T_{\max} &= C_u T_a \\
&= (1.4)(1.37 \text{ sec}) \\
&= 1.92 \text{ sec} \quad (1.9 \text{ sec})
\end{aligned}
$$

The upper limit is shorter than the period of vibration calculated by computer analysis, so it must be used.

The answer is (C).

Customary U.S. Solution

Use ASCE/SEI7 Eq. 12.8-7 to determine the approximate value for the period of vibration of the building. From ASCE/SEI7 Table 12.8-2, C_t is 0.0731 and x is 0.75 for an eccentrically based steel frame.

$$
\begin{aligned}
T_a &= C_t h_n^x \\
&= (0.03)(160 \text{ ft})^{0.75} \\
&= 1.35 \text{ sec}
\end{aligned}
$$

ASCE/SEI7 Sec. 12.8.2 defines the upper limit for the fundamental period as the product of C_u and T_a. From ASCE/SEI7 Table 12.8-1, C_u is 1.4 for buildings with S_{D1} values greater than 0.4. The upper-limit value is

$$
\begin{aligned}
T_{\max} &= C_u T_a \\
&= (1.4)(1.35 \text{ sec}) \\
&= 1.89 \text{ sec} \quad (1.9 \text{ sec})
\end{aligned}
$$

The upper limit is shorter than the period of vibration calculated by computer analysis, so it must be used.

The answer is (C).

Solutions
Practice Exam 2

56. The location of the earthquake is shown by the geographic position of its epicenter (point H) and its focal depth. The epicenter is the location on the earth's surface directly above the originating point of the earthquake (point I), which is also known as the hypocenter or focus. The focal depth as shown is the depth from the earth's surface to the hypocenter. Point G represents the closest distance to the fault plane projection.

The answer is (C).

57. The ASCE/SEI7 seismic design criteria and NEHRP provisions set minimum requirements that prevent loss of life, minimize loss of function, and preserve property. The specific intents are as follows.

From a small earthquake, buildings and structures should be free of damage.

From a moderate earthquake, buildings and structures may have some architectural (nonstructural) damage, but no structural damage.

From a severe earthquake, buildings and structures may sustain possible structural and nonstructural damage, but no structural collapse.

The answer is (A).

58. For proper seismic performance, a building must have a lateral force-resisting system that forms a load path between the foundation and all portions of the building. Redundancy is another characteristic that will improve performance in the event of an earthquake. ASCE/SEI7 Sec. 12.3.4 provides procedures to calculate the redundancy factor.

Structures that feature multiple load paths are called redundant. Redundancy is a measure of the number of alternate load paths that exist for primary structural elements, connections, and/or components such that if an element, and/or connection, and/or component fails, the capacity of alternate elements, connections, and/or components is available to satisfactorily resist the demand loads. As a result, the structure will remain laterally stable after the failure of any single element. Redundancy tends to mitigate high demand/capacity ratios. In structures that lack redundancy, all components must remain operative for the structure to maintain its lateral stability.

Rehabilitation techniques that enhance the redundancy of an existing structure include the addition of new lateral-load resisting elements or new systems to supplement existing weak or brittle systems. A multiple redundant stress path greatly increases the reliability of high-rise buildings.

The answer is (C).

59. ASCE/SEI7 Table 12.6-1 gives the permitted analytical procedures for designing structures. According to this table, the equivalent lateral-force procedure can be used in

1. buildings with occupancy categories I or II that have light-framed construction not exceeding three stories in height

2. other buildings with occupancy categories I or II not exceeding two stories in height

3. regular structures with $T < 3.5T_s$ and all structures of light frame construction

4. irregular structures with $T < 3.5T_s$ and that have only horizontal irregularities type 2, 3, 4, or 5 from ASCE/SEI7 Table 12.3-1, or vertical irregularities type 4, 5a, or 5b from ASCE/SEI7 Table 12.3-2

Structure I is an irregular structure because it has a torsional irregularity, which is a type 1a horizontal irregularity according to ASCE/SEI7 Table 12.3-1. Type 1a irregularities are not permitted to use the equivalent lateral-force procedure according to ASCE/SEI7 Table 12.6-1.

Structure II is a regular structure because it has no significant physical discontinuities in plan and vertical configuration, or in the lateral force-resisting systems. ASCE/SEI7 Table 12.6-1 states that the equivalent lateral-force procedure is permitted for regular structures with $T < 3.5T_s$. For structure II,

$$T_s = \frac{S_{D1}}{S_{DS}} = \frac{0.4}{0.833}$$
$$= 0.48 \text{ sec}$$
$$3.5T_s = (3.5)(0.48 \text{ sec})$$
$$= 1.68 \text{ sec}$$

$T = 2.0 > 3.5T_s$, therefore, the equivalent lateral-force procedure cannot be used for structure II.

The answer is (D).

60. Any civil engineering document that conveys an engineering recommendation or decision should bear the seal and signature of a registered civil engineer who either prepared the document or supervised its preparation. Examples of civil engineering documents are "Project Study Report," "Drainage Report," "Design Exception Fact Sheets," "Project Plan Sheets," "Project Report," "Project Specifications," "Material Report," "Special Provisions," and so on.

If the final civil engineering plans, specifications, reports, or documents have multiple pages, the signature and seal or stamp should appear on the original of the plans and on the original title sheet of the specifications, reports, or documents.

The answer is (D).

61. ASCE/SEI7 Sec. 15.4 provides information regarding the seismic design requirements for nonbuilding structures. Section 15.4.1 states that nonbuilding structures must be designed to resist minimum seismic lateral forces as given in Sec. 12.8, with some exceptions.

Because the monument is a rigid nonbuilding structure with a fundamental period less than 0.06 sec, the base shear for the monument is found using ASCE/SEI7 Eq. 15.4-5 [ASCE/SEI7 Sec. 15.4.2]. The importance factor for a risk category II structure is 1.0 [ASCE/SEI7 Table 1.5-2]. Because the structures are assumed to have the same weight, the weight can be disregarded. The base shear of the monument is

$$V = 0.30 S_{DS} W I_e = (0.30)(1.2) W (1.0)$$
$$= 0.36 W$$

Because the amusement structure is a rigid nonbuilding structure with a fundamental period greater than 0.06 sec, the base shear of the structure is found using Eq. 12.8-1. To find the base shear using this equation, the value of C_s must be known. For an amusement structure, the value of C_s is found using Eq. 12.8-2. Because amusement structures are nonbuilding structures that are not similar to buildings, and because the value of S_1 for the site is less than or equal to 0.6, the minimum value of C_s can be found using Eq. 15.4-2 [ASCE/SEI7 Sec. 15.4.1]. Because the fundamental period is less than any of the long-period transition periods given in ASCE/SEI7 Sec. 22, the maximum value of C_s can be found using Eq. 12.8-3 [ASCE/SEI7 Sec. 12.8.1.1]. For monuments, $R = 2$ [ASCE/SEI7 Table 15.4-2], and for risk category II buildings, $I_e = 1.0$ [ASCE/SEI7 Table 1.5-2].

The minimum value of C_s is

$$C_{s,\min} = \frac{0.8 S_1}{\dfrac{R}{I_e}} = \frac{(0.8)(0.6)}{\dfrac{2}{1.0}}$$
$$= 0.24$$

To find the maximum value of C_s, the value of S_{D1} must be found. Per ASCE/SEI7 Sec. 12.8.1.1, the value of S_{D1} is found using Eq. 11.4-4. This calculation requires the value of S_{M1}, which is found using Eq. 11.4-2. Per Table 11.4-2, the site coefficient F_v for site class C and $S_1 = 0.6$ is 1.4. The value of S_{M1} for the site is

$$S_{M1} = F_v S_1 = (1.4)(0.6)$$
$$= 0.84$$

The value of S_{D1} for the site is

$$S_{D1} = \frac{2}{3} S_{M1} = \left(\frac{2}{3}\right)(0.84)$$
$$= 0.56$$

The maximum value of C_s is

$$C_{s,\max} = \frac{S_{D1}}{T\left(\dfrac{R}{I_e}\right)} = \frac{0.56}{(0.05 \text{ sec})\left(\dfrac{2}{1.0}\right)}$$
$$= 5.6$$

The value of C_s is

$$C_s = \frac{S_{DS}}{\dfrac{R}{I_e}} = \frac{1.2}{\dfrac{2}{1.0}} = 0.6 \quad \text{[governs]}$$

The base shear of the monument, disregarding the weight of the structure, is

$$V = C_s W = 0.6 W$$

The monument has the smaller base shear.

The answer is (A).

62. Unreinforced masonry (URM) is perhaps the most dangerous type of construction in areas of high seismicity, such as California. The Federal Emergency Management Agency recognizes URM as one of the structure types most prone to failure during an earthquake. URM structures are made from brick, hollow clay tile, stone, or concrete blocks that are not strengthened by additional steel rods or bracings. The masonry is usually held together with weak mortar, making it unable to resist lateral forces. Wall and roof anchorages also tend to be inadequate, which allows roofs and floors to separate from the structure's walls and collapse. While not all URM buildings will collapse during a significant earthquake, most will have some degree of life-threatening failure. In California, the construction of unreinforced masonry buildings was banned after the 1933 Long Beach earthquake, which killed over 100 people.

Tilt-up construction is increasingly common due to its efficiency and ability to reduce construction costs from that of traditional masonry construction. Tilt-up construction refers to structures where panels are fabricated onsite in a horizontal mold, then "tilted up" to form vertical walls. OSHA Standards for Construction require that during construction, tilt-up wall panels have adequate temporary support (e.g., metal straps between concrete walls and the roof framing system) to prevent collapse or overturning until permanent connections are completed. Tilt-up structures that experience earthquake damage are usually poorly constructed with inadequate ties between the roof and the walls, resulting in separation.

An infill wall is often used to increase the lateral strength and rigidity of a structure by filling space between the beams and columns with masonry or cast-in-place concrete. An infill wall of a moment frame structure becomes a part of the lateral-force resisting system (i.e., acts as a shear wall in conjunction with the frame), and during an earthquake, these walls can become damaged or fail. However, in a well-designed structure, the frame is unlikely to collapse.

Prefabricated metal buildings have generally performed well during an earthquake. They are normally one-story, relatively lightweight structures with steel framing, metal decking and siding, and roofs with horizontal rod bracing. Lateral forces are resisted by moment-frame action in one direction and rod bracing in the other.

The answer is (C).

63. Nonstructural architectural elements can be damaged in the event of an earthquake, and some of this damage may become a life-threatening hazard. The two principal causes of nonstructural architectural damage are (1) insufficient anchorage capacity between structural and architectural elements, and (2) differential seismic displacement.

As an example, differential seismic displacement (i.e., drift) between stories of a building can cause damage to windows. Architectural appendages (such as cornices, parapets, and spandrels) or architectural cladding (such as a granite veneer) with insufficient anchorage capacity are susceptible to damage and may spall. Precast concrete cladding, with or without stone facing, is heavy. The steel connections holding this type of cladding onto the structure should be strong enough to allow the building to move in an earthquake without failing. In addition, gaps or joints between cladding units should be large enough and should be positioned in the proper places to accommodate building movement.

Architectural parapets, if not adequately braced to roof framing, can be a life-threatening hazard since they can fall in an earthquake. There are different techniques that can be used to strengthen deficient parapets:

(1) parapets can be reduced in height so that the parapet dead load will resist uplift from out-of-plane seismic forces, (2) parapets can be strengthened with a new concrete overlay (shotcrete), and/or (3) parapets can be braced back to roof framing.

Requirements for building separations are outlined in ASCE/SEI7 Sec. 12.12.3. When buildings are constructed too close together, they may pound each other under the influence of the swaying caused by an earthquake. While this may also result in architectural damage, the more serious result can be damage to structural elements and collapse of the building.

The answer is (B).

64. *SI Solution*

The base shear distributes to each story according to ASCE/SEI7 Eq. 12.8-11 and Eq. 12.8-12.

$$F_x = C_{vx}V \qquad \text{[ASCE/SEI7 Eq. 12.8-11]}$$

$$C_{vx} = \frac{w_x h_x^k}{\displaystyle\sum_{i=1}^{n} w_i h_i^k} \qquad \text{[ASCE/SEI7 Eq. 12.8-12]}$$

Since the fundamental time period is 0.6 s, k is determined by linear interpolation. For structures having a period of 0.5 s or less, $k = 1$. For structures having a period greater than 2.5 s, $k = 2$. Therefore,

$$k = 1 + \left(\frac{0.6 \text{ s} - 0.5 \text{ s}}{2.5 \text{ s} - 0.5 \text{ s}}\right)(2-1) = 1.05 \text{ s}$$

A table is the easiest way to set up the data for calculating and recording story shears.

level i	h_x (m)	w_x (kN)	$w_x h_x^k$	$\dfrac{w_x h_x^k}{\sum w_x h_x^k}$	F_x (kN)
5	18.5	2200	47 093	0.340	
4	14.8	2200	37 256	0.269	140 (given)
3	11.1	2200	27 543	0.199	
2	7.4	2200	17 994	0.130	
1	3.7	2200	8690	0.063	
		$\sum w_x h_x^k$	138 576		

Given that $F_4 = 140$ kN,

$$C_{v4} = 0.269 = \frac{F_4}{V} = \frac{140 \text{ kN}}{V}$$

$$V = \frac{140 \text{ kN}}{0.269}$$
$$= 520.4 \text{ kN} \quad (520 \text{ kN})$$

The answer is (B).

Customary U.S. Solution

The base shear distributes to each story according to ASCE/SEI7 Eq. 12.8-11 and Eq. 12.8-12.

$$F_x = C_{vx}V \qquad \text{[ASCE/SEI7 Eq. 12.8-11]}$$

$$C_{vx} = \frac{w_x h_x^k}{\displaystyle\sum_{i=1}^{n} w_i h_i^k} \qquad \text{[ASCE/SEI7 Eq. 12.8-12]}$$

Since the fundamental time period is 0.6 sec, k is determined by linear interpolation. For structures having a period of 0.5 sec or less, $k = 1$. For structures having a period greater than 2.5 sec, $k = 2$. Therefore,

$$k = 1 + \left(\frac{0.6 \text{ sec} - 0.5 \text{ sec}}{2.5 \text{ sec} - 0.5 \text{ sec}}\right)(2-1)$$
$$= 1.05 \text{ sec}$$

A table is the easiest way to set up the data for calculating and recording story shears.

level i	h_x (ft)	w_x (kips)	$w_x h_x^k$	$\dfrac{w_x h_x^k}{\sum w_x h_x^k}$	F_x (kips)
5	60	500	36,815	0.340	
4	48	500	29,125	0.268	32 (given)
3	36	500	21,532	0.199	
2	24	500	14,067	0.130	
1	12	500	6794	0.063	
			$\sum w_x h_x^k$ 108,333		

Given that $F_4 = 32$ kips,

$$C_{v4} = 0.268 = \frac{F_4}{V} = \frac{32 \text{ kips}}{V}$$

$$V = \frac{32 \text{ kips}}{0.268}$$
$$= 119.4 \text{ kips} \quad (120 \text{ kips})$$

The answer is (B).

65. The Alquist-Priolo Earthquake Fault Zoning Act addresses the hazard of surface fault rupture that occurs when movement on a fault deep within the earth breaks through to the surface. It does not address other earthquake hazards, such as ground shaking or liquefaction. The law requires that the State Geologist establish regulatory zones around the surface traces of active faults and issue appropriate maps.

The main purpose of this act is to prevent the construction of buildings used for human occupancy on the surface trace of active faults. This act exempts the construction of single-family wood-frame and steel-frame dwellings up to two stories that are not part of a development of four units or more. Local agencies regulate most development projects within the zones, and they can be more restrictive than the state law requires.

The construction of buildings that store explosive and/or hazardous materials should meet seismic design requirements if local agencies permit their construction in their jurisdictions.

The answer is (A).

66. At points of discontinuity in the plan, or where the vertical resisting elements (i.e., shear walls or braced frames) are not provided because of windows or doors, struts (collectors or ties) collect and drag the horizontal diaphragm shear to the supporting vertical resisting elements. The perimeter members of a horizontal roof diaphragm that are parallel to the applied lateral force are the struts. Members F and H are struts.

Chords are boundary members of a diaphragm that are perpendicular to the direction of the lateral load. Their function is to carry moment and provide all the resistance to the flexural stresses. Chord elements for north-south earthquake loading are located along the north and south walls.

The members that serve as the struts in the direction of the applied lateral force are often the same members that function as the chords for the lateral force in the other direction.

The answer is (B).

67. The force is distributed to the columns in proportion to their rigidities (stiffnesses). The composite stiffness is the sum of the individual column stiffnesses.

$$K_T = K_{\mathrm{I}} + K_{\mathrm{II}} \qquad \text{[Eq. 1]}$$

The distributed shear stresses in the first and second columns, respectively, are

$$V_{\mathrm{I}} = F\left(\frac{K_{\mathrm{I}}}{K_T}\right) \qquad \text{[Eq. 2]}$$

$$V_{\mathrm{II}} = F\left(\frac{K_{\mathrm{II}}}{K_T}\right) \qquad \text{[Eq. 3]}$$

Column I is fixed at the top and free at the bottom (i.e., cantilever action). The stiffness of a cantilever beam is

$$K_{\mathrm{I}} = \frac{3EI}{h^3}$$

Column II is rigid at both ends, and the stiffness for this column is

$$K_{\mathrm{II}} = \frac{12EI}{h^3}$$

In Eq. 1, substitute for K_I and K_{II}.

$$K_T = \frac{3EI}{h^3} + \frac{12EI}{h^3} = 15EI/h^3$$

In Eq. 2 and Eq. 3, substitute for K_T, K_I, and K_{II}, then solve for V_I and V_{II}.

$$V_I = F\left(\frac{\frac{3EI}{h^3}}{\frac{15EI}{h^3}}\right) = \frac{1}{5}F$$

$$V_{II} = F\left(\frac{\frac{12EI}{h^3}}{\frac{15EI}{h^3}}\right) = \frac{4}{5}F$$

Therefore,

$$\frac{V_I}{V_{II}} = \frac{\frac{1}{5}F}{\frac{4}{5}F}$$

$$\frac{4}{5}V_I = \frac{1}{5}V_{II}$$

$$V_I = \frac{1}{4}V_{II}$$

The answer is (B).

68. Steel braced frames are earthquake-resistant structures. There are two types of steel braced frames: concentrically braced frames and eccentrically braced frames. ASCE/SEI7 Sec. 11.2 gives the definitions for both.

Concentrically braced frames are braced frames whose members are subjected primarily to axial forces. This is accomplished when the centerlines of the beam or column members coincide with the ends of the frame members. Structures I (X-bracing), II (inverted V-bracing), and IV (K-bracing) are concentrically braced frames.

Eccentrically braced frames are diagonal steel-braced frames in which at least one end of each bracing member connects to a beam a short distance (link beam) from a beam-to-column connection, or from another beam-to-brace connection. Link beams increase the ductility of the frame. Structure III is an eccentrically braced frame.

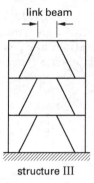

structure III

The answer is (D).

69. ASCE/SEI7 Sec. 12.3.1.2 describes rigid diaphragms, which include concrete slabs or concrete-filled metal decks that have span-to-depth ratios no greater than three and no horizontal irregularities.

ASCE/SEI7 Sec. 12.3 describes flexible and rigid diaphragms. According to ASCE/SEI7 Sec. 12.3.1.1, flexible diaphragms can be constructed of wood structural panels in which the vertical elements are steel (or composite steel) and concrete braced frames; or, are concrete, masonry, steel, or composite shear walls. Untopped steel decks in one- and two-family residential buildings of light-frame construction may also be idealized as flexible.

The answer is (A).

70. The seismometer records the amplitude of an earthquake. The Richter magnitude, M, for the earthquake is determined from the logarithm of the recorded displacement amplitude. Since the Richter magnitude is a logarithmic scale, each whole number increase in magnitude corresponds to a 10 time increase in measured amplitude. Thus, an earthquake of magnitude 7.0 represents a 100 time increase in measured amplitude over an earthquake of magnitude 5.0.

The answer is (B).

71. High-rise and bridge structures should be designed and detailed by professional civil or structural engineers who are fully qualified and have design experience in the following areas: construction materials, determination of lateral forces, selection of structural systems, selection of foundation systems, application of code requirements, and multistory structures. The reason is that the stability of a structure is dependent upon the interaction of the individual structural components as well as the structure as a whole.

Following the 1933 Long Beach earthquake, the Field Act gave school design approval to the Division of the State Architect. As required by the Division of the State Architect, State Department of General Services, registered structural engineers must sign and stamp engineering design plans for schools in California.

The answer is (D).

72. Use ASCE/SEI7 Sec. 13.3 and Eq. 13.3-1, Eq. 13.3-2, and Eq. 13.3-3 to determine the design lateral seismic force on structural elements, nonstructural components, and equipment supported by structures.

$$F_p = \left(\frac{0.4 a_p S_{DS} W_p}{\dfrac{R_p}{I_p}}\right)\left(1 + 2\frac{z}{h}\right)$$

[ASCE/SEI7 Eq. 13.3-1]

$$F_{p,\max} = 1.6 S_{DS} I_p W_p \qquad \text{[ASCE/SEI7 Eq. 13.3-2]}$$

$$F_{p,\min} = 0.3 S_{DS} I_p W_p \qquad \text{[ASCE/SEI7 Eq. 13.3-3]}$$

From ASCE/SEI7 Table 13.6-1, item 2, the values of a_p and R_p are 1.0 and 2.5, respectively. ASCE/SEI7 Sec. 13.1.3 gives I_p as 1.5 for components that are required to function for life-safety purposes after an earthquake, including fire protection sprinkler systems. Since the connection of the tank is at the roof of the building, $z/h = 1.0$.

$$F_p = \left(\frac{(0.4)(1.0)(1.2)\,W}{\dfrac{2.5}{1.5}}\right)\bigl(1 + (2)(1)\bigr) = 0.86\,W$$

$$F_{p,\max} = (1.6)(1.2)(1.5)\,W = 2.88\,W$$

$$F_{p,\min} = (0.3)(1.2)(1.5)\,W = 0.54\,W$$

The correct value is $0.86\,W\,(0.90\,W)$.

The answer is (C).

73. ASCE/SEI7 Table 12.2-1 does not permit ordinary concentrically braced frames (OCBF) in seismic design category D buildings, unless either the building is 35 ft or less, or it is a single-story building less than 60 ft with a roof dead load less than 20 lbf/ft². Since the building satisfies neither of the OCBF conditions, option I is incorrect.

ASCE/SEI7 Table 12.2-1 does not permit the use of an ordinary steel moment frame (OMF) in structures assigned to seismic design category D. However, the table's footnote i refers to ASCE/SEI7 Sec. 12.2.5.6, which clarifies that OMFs *are* permitted if the given conditions are met.

For single-story buildings, OMFs are permitted up to a height of 65 ft (20 m) where the dead load supported by and tributary to the roof does not exceed 20 lbf/ft² (0.96 kN/m²). Additionally, the dead load of the exterior wall supported by the OMF cannot exceed 20 lbf/ft² (0.96 kN/m²).

For multi-story buildings framed as light frame construction, up to a height of 35 ft (10.6 m), OMFs are permitted where neither the dead load of the roof nor the floor supported by it exceeds 20 lbf/ft² (0.96 kN/m²). Additionally, the dead load of the exterior wall supported by the OMF cannot exceed 20 lbf/ft² (0.96 kN/m²).

Because the building does not satisfy any of the above conditions, option II is incorrect.

Eccentrically braced frames (EBF) are allowed for buildings up to 160 ft and assigned to seismic design category D. Requirements for eccentrically braced frames are given in ANSI/AISC 341 Sec. F3. The building satisfies the requirements for an EBF frame; therefore, option III is correct.

The answer is (B).

74. The center of mass is at the center of the diaphragm, while the center of rigidity is at the centerline in the east-west direction, located near wall B in the north-south direction. The earthquake shear force in north-south loading is resisted directly by walls A and C. These walls share the load equally and have the same amounts of displacement because they have equal rigidity. The centers of mass and rigidity coincide for this loading direction, so there is no eccentricity and no torsion. Note that ASCE/SEI7 Sec. 12.8.4.2 stipulates that an accidental eccentricity of 5% of the building length perpendicular to the force must be considered in determining the torsional design moment.

The answer is (B).

75. ASCE/SEI7 Table 12.3-2 defines five types of structural irregularities and lists their corresponding appropriate reference sections. Vertical irregularities may adversely affect a structure's seismic resistance, so ASCE/SEI7 has developed additional regulations and penalties for these structures.

An abrupt change in story deflection usually reveals a stiffness irregularity (soft story). Using ASCE/SEI7 Table 12.3-2 and a visual observation of the plotted story deflection, the soft story vertical irregularity is identified by the building's abrupt change shown in the second story.

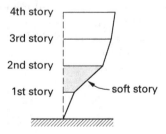

A stiffness soft story irregularity exists where there is a story whose lateral stiffness is less than 70% of that in the story above, or less than 80% of the average stiffness of the three stories above. A stiffness-extreme soft story irregularity exists where there is a story whose lateral stiffness is less than 60% of that in the story above, or less than 70% of the average stiffness of the three stories above.

Examples of soft stories are commercial buildings with open fronts or sides at ground level, and stores, hotels, and office buildings with especially tall first stories.

The answer is (B).

76. *SI Solution*

The chord force is the bending moment divided by the diaphragm depth.

$$C = \frac{M}{b} \qquad \text{[Eq. 1]}$$

To find the bending moment at the intersection of lines Z and 1, develop the shear and bending moment diagrams along the south wall.

The diaphragm shear force in each parallel wall is

$$V = \frac{wL}{2} = \frac{\left(3600 \ \frac{N}{m}\right)(24 \ m)}{2} = 43\,200 \ N$$

The shear and bending moment diagrams across the length of the chord are as shown.

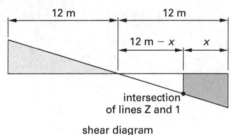

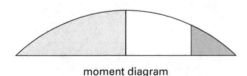

$$\frac{43\,200 \ N}{12 \ m} = \frac{V}{x}$$
$$V = (3600 \ N/m)x$$

The moment at line 1 is calculated as the area under the shear diagram.

$$M = \left(43\,200 \ N + \left(3600 \ \frac{N}{m}\right)x\right)\left(\frac{x}{2}\right)$$
$$= (21\,600 \ N/m)x + (1800 \ N/m)x^2$$

In Eq. 1, substitute for $C = 16\,500$ N, $b = 12$ m, and $M = (21\,600 \ N/m)x + (1800 \ N/m)x^2$. Solve for x.

$$16\,500 \ N = \frac{\left(21\,600 \ \frac{N}{m}\right)x + \left(1800 \ \frac{N}{m}\right)x^2}{12 \ m}$$

$$x^2 + (12 \ m)x - 110 \ m = 0$$
$$x = 6.1 \ m$$

Therefore, the total length of the south wall is

$$L = 12 \ m + (12 \ m - 6.1 \ m) = 17.9 \ m \quad (18 \ m)$$

The answer is (C).

Customary U.S. Solution

The chord force is the bending moment divided by the diaphragm depth.

$$C = \frac{M}{b} \qquad \text{[Eq. 1]}$$

To find the bending moment at the intersection of lines Z and 1, develop the shear and bending moment diagrams along the south wall.

The diaphragm shear force in each parallel wall is

$$V = \frac{wL}{2} = \frac{\left(250 \ \frac{lbf}{ft}\right)(80 \ ft)}{2} = 10,000 \ lbf$$

The shear and bending moment diagrams across the length of the chord are as shown.

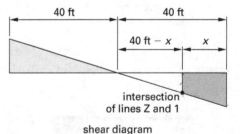

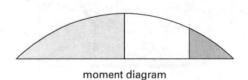

$$\frac{10,000 \ lbf}{40 \ ft} = \frac{V}{x}$$
$$V = (250 \ lbf/ft)x$$

The moment at line 1 is calculated as the area under the shear diagram.

$$M = \left(10{,}000 \text{ lbf} + \left(250 \ \frac{\text{lbf}}{\text{ft}}\right)x\right)\left(\frac{x}{2}\right)$$
$$= (5000 \text{ lbf})x + (125 \text{ lbf/ft})x^2$$

In Eq. 1, substitute for $C = 3750$ lbf, $b = 40$ ft, and $M = (5000 \text{ lbf})x + (125 \text{ lbf/ft})x^2$. Solve for x.

$$3750 \text{ lbf} = \frac{(5000 \text{ lbf})x + \left(125 \ \frac{\text{lbf}}{\text{ft}}\right)x^2}{40 \text{ ft}}$$

$$x^2 + (40 \text{ ft})x - 1200 \text{ ft} = 0$$
$$x = 20 \text{ ft}$$

Therefore, the total length of the south wall is

$$L = 40 \text{ ft} + (40 \text{ ft} - 20 \text{ ft}) = 60 \text{ ft}$$

The answer is (C).

77. ASCE/SEI7 Sec. 21.2.3 states the site-specific MCE_R spectral response acceleration is the lesser of the spectral response accelerations from the probabilistic MCE_R [ASCE/SEI7 Sec. 21.2.1] and the deterministic MCE_R [ASCE/SEI7 Sec. 21.2.2].

Probabilistic MCE_R spectral response accelerations are represented by a 5% damped acceleration response spectrum that has a 2% exceedence probability within a 50 yr period. Deterministic MCE_R response accelerations are calculated as 150% of the largest median 5% damped spectral response acceleration, computed at a period for the characteristic earthquakes on all known active faults within the region.

The answer is (C).

78. *SI Solution*

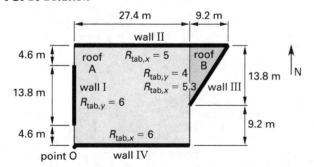

point O   wall IV

To find the location of the building center of mass in the x-direction, the dead loads of the roof and walls should be calculated.

For roof A,

$$D = \frac{(27.4 \text{ m})(23 \text{ m})\left(1200 \ \dfrac{\text{N}}{\text{m}^2}\right)}{1000 \ \dfrac{\text{N}}{\text{kN}}}$$
$$= 756.2 \text{ kN}$$

For roof B,

$$D = \frac{(9.2 \text{ m})(13.8 \text{ m})\left(\dfrac{1}{2}\right)\left(1200 \ \dfrac{\text{N}}{\text{m}^2}\right)}{1000 \ \dfrac{\text{N}}{\text{kN}}}$$
$$= 76.2 \text{ kN}$$

For all walls,

$$D = \frac{(3.7 \text{ m})\left(2200 \ \dfrac{\text{N}}{\text{m}^2}\right)}{1000 \ \dfrac{\text{N}}{\text{kN}}}$$
$$= 8.14 \text{ kN/m}$$

For wall I,

$$D = (13.8 \text{ m})\left(8.14 \ \frac{\text{kN}}{\text{m}}\right)$$
$$= 112.3 \text{ kN}$$

For wall II,

$$D = (36.6 \text{ m})\left(8.14 \ \frac{\text{kN}}{\text{m}}\right)$$
$$= 297.9 \text{ kN}$$

For wall III,

$$L = \sqrt{(13.8 \text{ m})^2 + (9.2 \text{ m})^2}$$
$$= 16.59 \text{ m}$$
$$D = (16.59 \text{ m})\left(8.14 \ \frac{\text{kN}}{\text{m}}\right)$$
$$= 135.0 \text{ kN}$$

For wall IV,

$$D = (27.4 \text{ m})\left(8.14 \ \frac{\text{kN}}{\text{m}}\right) = 223.0 \text{ kN}$$

The easiest way to set up the data for determining the building center of mass is to use a table. Distances are measured from the southwest corner (point O).

	D (kN)	x (m)	Dx (kN·m)
roof A	756.2	13.7	10 360
B	76.2	30.5	2324
wall I	112.3	0	0
II	297.9	18.3	5452
III	135.0	32.0	4320
IV	223.0	13.7	3055
total	$\sum D_i$ 1600.6		$\sum D_i x_i$ 25 511

$$\bar{x} = \frac{\sum D_i x_i}{\sum D_i} = \frac{25\,511 \text{ kN·m}}{1600.6 \text{ kN}} = 15.94 \text{ m} \quad (16 \text{ m})$$

The easiest way to determine the building center of rigidity is to use a table. Distances are measured from the southwest corner (point O). Tabulated rigidities in the x-direction, $R_{\text{tab},x}$, are not needed for this calculation.

wall	$R_{\text{tab},x}$	$R_{\text{tab},y}$	x (m)
I	0	6	0
II	5	0	18.3
III	5.3	4	32.0
IV	6	0	13.7

To determine the location of the building center of rigidity in the x-direction, consider the loading in the north-south direction. Omit the weak walls (i.e., walls II and IV); they are perpendicular to the direction of loading (only parallel walls can resist seismic shear forces).

$$\bar{x}_R = \frac{\sum R_{\text{tab},i} x_i}{\sum R_{\text{tab},i}} = \frac{(6)(0 \text{ m}) + (4)(32.0 \text{ m})}{6 + 4}$$
$$= 12.8 \text{ m} \quad (13 \text{ m})$$

The answer is (C).

Customary U.S. Solution

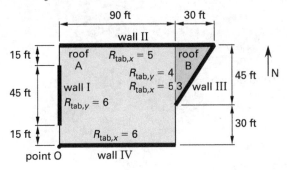

To find the location of the building center of mass in the x-direction, the dead loads of the roof and walls should be calculated.

For roof A,

$$D = \frac{(90 \text{ ft})(75 \text{ ft})\left(25 \ \dfrac{\text{lbf}}{\text{ft}^2}\right)}{1000 \ \dfrac{\text{lbf}}{\text{kip}}}$$
$$= 168.8 \text{ kips}$$

For roof B,

$$D = \frac{(30 \text{ ft})(45 \text{ ft})\left(\dfrac{1}{2}\right)\left(25 \ \dfrac{\text{lbf}}{\text{ft}^2}\right)}{1000 \ \dfrac{\text{lbf}}{\text{kip}}}$$
$$= 16.9 \text{ kips}$$

For all walls,

$$D = \frac{(12 \text{ ft})\left(45 \ \dfrac{\text{lbf}}{\text{ft}^2}\right)}{1000 \ \dfrac{\text{lbf}}{\text{kip}}}$$
$$= 0.54 \text{ kip/ft}$$

For wall I,

$$D = (45 \text{ ft})\left(0.54 \ \dfrac{\text{kip}}{\text{ft}}\right)$$
$$= 24.3 \text{ kips}$$

For wall II,

$$D = (120 \text{ ft})\left(0.54 \ \dfrac{\text{kip}}{\text{ft}}\right)$$
$$= 64.8 \text{ kips}$$

For wall III,

$$L = \sqrt{(45 \text{ ft})^2 + (30 \text{ ft})^2}$$
$$= 54 \text{ ft}$$
$$D = (54 \text{ ft})\left(0.54 \ \dfrac{\text{kip}}{\text{ft}}\right)$$
$$= 29.2 \text{ kips}$$

For wall IV,

$$D = (90 \text{ ft})\left(0.54 \ \dfrac{\text{kip}}{\text{ft}}\right)$$
$$= 48.6 \text{ kips}$$

The easiest way to set up the data for determining the building center of mass is to use a table. Distances are measured from the southwest corner (point O).

	D (kips)	x (ft)	Dx (ft-kips)
roof A	168.8	45	7596.0
B	16.9	100	1690.0
wall I	24.3	0	0
II	64.8	60	3888.0
III	29.2	105	3066.0
IV	48.6	45	2187.0
total	$\sum D_i$		$\sum D_i x_i$
	352.6		18,427.0

$$\bar{x} = \frac{\sum D_i x_i}{\sum D_i} = \frac{18,427 \text{ ft-kips}}{352.6 \text{ kips}}$$
$$= 52.26 \text{ ft} \quad (52 \text{ ft})$$

The easiest way to determine the building center of rigidity is to use a table. Distances are measured from the southwest corner (point O). Tabulated rigidities in the x-direction, $R_{\text{tab},x}$, are not needed for this calculation.

wall	$R_{\text{tab},x}$	$R_{\text{tab},y}$	x (ft)
I	0	6	0
II	5	0	60
III	5.3	4	105
IV	6	0	45

To determine the location of the building center of rigidity in the x-direction, consider the loading in the north-south direction. Omit the weak walls (i.e., walls II and IV); they are perpendicular to the direction of loading (only parallel walls can resist seismic shear forces).

$$\bar{x}_R = \frac{\sum R_{\text{tab},i} x_i}{\sum R_{\text{tab},i}} = \frac{(6)(0 \text{ ft}) + (4)(105 \text{ ft})}{6 + 4}$$
$$= 42 \text{ ft}$$

The answer is (C).

79. Wood-frame structures are normally constructed with flexible diaphragms, which are relatively thin structural elements supported by the shear walls. Shear walls comprise the vertical elements in the lateral-force resisting system. Flexible diaphragms distribute lateral forces to shear walls in proportion to their tributary areas, and the shear walls transfer the lateral loads into the foundation.

Rigid diaphragms transfer torsion to the vertical resisting elements, but flexible diaphragms do not.

The answer is (A).

80. *SI Solution*

The eccentricity is the distance between the centers of mass and rigidity, measured in the direction perpendicular to the lateral force. For the north-south loading, the eccentricity is

$$e = \bar{x}_R - \bar{x} = 13.7 \text{ m} - 9.8 \text{ m}$$
$$= 3.9$$

According to ASCE/SEI7 Sec. 12.8.4.2, accidental eccentricity, e_a, is taken as 5% of the building dimension in each direction perpendicular to the lateral force. For north-south loading,

$$e_a = \pm(0.05)(24.4 \text{ m})$$
$$= \pm 1.2 \text{ m}$$

The total design eccentricity is

$$e = e + e_a = 3.9 \text{ m} \pm 1.2 \text{ m}$$
$$= 5.1 \text{ m and } 2.7 \text{ m}$$

Use $e = 5.1$ m.

The answer is (C).

Customary U.S. Solution

The eccentricity is the distance between the centers of mass and rigidity, measured in the direction perpendicular to the lateral force. For the north-south loading, the eccentricity is

$$e = \bar{x}_R - \bar{x} = 45 \text{ ft} - 32 \text{ ft}$$
$$= 13 \text{ ft}$$

According to ASCE/SEI7 Sec. 12.8.4.2, accidental eccentricity, e_a, is taken as 5% of the building dimension in each direction perpendicular to the lateral force. For north-south loading,

$$e_a = \pm(0.05)(80 \text{ ft})$$
$$= \pm 4 \text{ ft}$$

The total design eccentricity is

$$e = e + e_a = 13 \text{ ft} \pm 4 \text{ ft} = 17 \text{ ft and } 9 \text{ ft}$$

Use $e = 17$ ft.

The answer is (C).

81. Based on the Professional Engineers Act and the Board Rules and Regulations Relating to the Practices of Professional Engineering and Professional Land Surveying [Sec. 404.1(b)], to evaluate whether a professional civil engineer is in responsible charge of his or her engineering work, the engineer must be capable of answering questions relevant to the engineering decisions personally made during the project when asked by equally qualified engineers. The professional engineer should also possess sufficient knowledge of the project. It is not necessary for the professional engineer to defend his or her decisions in an adversarial situation.

The answer is (B).

82. Rigid diaphragms distribute lateral loads to resisting vertical elements (i.e., shear walls) in proportion to their relative rigidities. The walls perpendicular to the direction of the applied lateral load resist the bending moment and also contribute inertia load to the diaphragm shear. Walls parallel to the lateral force resist seismic shear forces. Walls I, III, and IV are perpendicular to the lateral force. Of these, wall III has the smallest tabulated rigidity and, therefore, the smallest relative rigidity. Therefore, wall III resists less of the lateral force than walls I and IV.

The answer is (B).

83. A fault is a fracture in the earth's crust along which two crustal blocks have slipped with respect to each other. Most of the major faults in California are vertical or near-vertical breaks. Movement along these breaks is predominantly horizontal in the northerly or northwesterly direction. Major faults in California are the San Andreas, Hayward, Calaveras, Rodgers Creek, Green Valley, Concord, Antioch, and San Gregorio faults. Earthquakes in California are relatively shallow; in other words, they have shallow focal depths.

The San Andreas fault is the major fault that cuts through rocks of the coastal region. Many smaller faults branch from and join this fault. Movement along the San Andreas fault is predominantly right-lateral, which means that if a person stands on one side of the fault facing it, the block on the other side moves to the right. A segment of the San Andreas fault slipped on October 17, 1989, causing the magnitude 7.1 Loma Prieta earthquake. During this earthquake, a 25 mi (40 km) segment of the San Andreas fault southwest of San Jose slipped about 7 ft (2.1 m).

The answer is (B).

84. Based on the Professional Engineers Act, Business and Professions Code Sec. 6737.1, architects, civil engineers, and structural engineers can sign final calculations of dead plus live plus seismic combinations of loads for a two-story, single-family wood-frame residential building.

The answer is (D).

85. *SI Solution*

The design base shear (i.e., the total force) can be obtained from

$$V = mS_a$$

From the problem table, and with the natural period, T, of 0.3 s, the response spectra is $S_a = 2.0$ g.

$$V = mS_a = (11\,000 \text{ kg})\left(9.81 \ \frac{\text{m}}{\text{s}^2 \cdot \text{g}}\right)(2.0 \text{ g})$$

$$= 215\,820 \text{ N}$$

$$\Delta = V\left(\frac{1}{k}\right) = (215\,820 \text{ N})\left(\frac{(1)\left(1000 \ \frac{\text{mm}}{\text{m}}\right)}{5.4 \times 10^6 \ \frac{\text{N}}{\text{m}}}\right)$$

$$= 40 \text{ mm}$$

The answer is (B).

Customary U.S. Solution

The design base shear (i.e., the total force) can be obtained from

$$V = mS_a$$

From the problem table, and with the natural period, T, of 0.3 sec, the response spectra is $S_a = 2.0$ g.

$$V = mS_a = (25{,}000 \text{ lbf})(2.0 \text{ g})$$

$$= 50{,}000 \text{ lbf}$$

$$\Delta = V\left(\frac{1}{k}\right) = (50{,}000 \text{ lbf})\left(\frac{1}{30{,}000 \ \frac{\text{lbf}}{\text{in}}}\right)$$

$$= 1.7 \text{ in}$$

The answer is (B).

86. *SI Solution*

The absolute rigidity of the entire wall is

$$R_{\text{abs,entire wall}} = R_{\text{abs,masonry wall}} + R_{\text{abs,steel frame}}$$
$$+ R_{\text{abs,concrete wall}}$$
$$= 40 \text{ kN/mm}$$

The masonry shear wall, steel frame, and concrete shear wall resist the lateral load in proportion to their relative rigidities. The total lateral load is

$$V_{total} = V_{masonry\,wall} + V_{steel\,frame} + V_{concrete\,wall}$$
$$= 180 \text{ kN} + 70 \text{ kN} + 90 \text{ kN}$$
$$= 340 \text{ kN}$$

For the masonry shear wall,

$$V_{masonry\,wall} = \left(\frac{R_{abs,masonry\,wall}}{R_{abs,entire\,wall}} \right) V_{total}$$

$$R_{abs,masonry\,wall} = \frac{V_{masonry\,wall} R_{abs,entire\,wall}}{V_{total}}$$

$$= \frac{(180 \text{ kN})\left(40 \ \frac{\text{kN}}{\text{mm}} \right)}{340 \text{ kN}}$$

$$= 21.2 \text{ kN/mm}$$

For the steel frame,

$$V_{steel\,frame} = \left(\frac{R_{abs,steel\,frame}}{R_{abs,entire\,wall}} \right) V_{total}$$

$$R_{abs,steel\,frame} = \frac{V_{steel\,frame} R_{abs,entire\,wall}}{V_{total}}$$

$$= \frac{(70 \text{ kN})\left(40 \ \frac{\text{kN}}{\text{mm}} \right)}{340 \text{ kN}}$$

$$= 8.24 \text{ kN/mm}$$

For the concrete wall,

$$V_{concrete\,wall} = \left(\frac{R_{abs,steel\,frame}}{R_{abs,entire\,wall}} \right) V_{total}$$

$$R_{abs,concrete\,wall} = \frac{V_{concrete\,wall} R_{abs,entire\,wall}}{V_{total}}$$

$$= \frac{(90 \text{ kN})\left(40 \ \frac{\text{kN}}{\text{mm}} \right)}{340 \text{ kN}}$$

$$= 10.6 \text{ kN/mm}$$

The masonry wall has the largest absolute rigidity of 21.2 kN/mm. (Statement II is true.)

The relative rigidities of the two shear walls and the steel frame are

$$R_{rel,i} = \frac{R_{abs,i}}{\sum_i R_{abs,i}} = \frac{R_{abs,i}}{R_{abs,entire\,wall}}$$

$$R_{rel,masonry\,wall} = \frac{R_{abs,masonry\,wall}}{R_{abs,entire\,wall}} = \frac{21.2 \ \frac{\text{kN}}{\text{mm}}}{40 \ \frac{\text{kN}}{\text{mm}}}$$

$$= 0.529$$

$$R_{rel,steel\,frame} = \frac{R_{abs,steel\,frame}}{R_{abs,entire\,wall}} = \frac{8.24 \ \frac{\text{kN}}{\text{mm}}}{40 \ \frac{\text{kN}}{\text{mm}}}$$

$$= 0.206$$

$$R_{rel,concrete\,wall} = \frac{R_{abs,concrete\,wall}}{R_{abs,entire\,wall}} = \frac{10.6 \ \frac{\text{kN}}{\text{mm}}}{40 \ \frac{\text{kN}}{\text{mm}}}$$

$$= 0.265$$

$R_{rel,steel\,frame} < R_{rel,concrete\,wall} < R_{rel,masonry\,wall}$, so the steelframe does not have the largest relative rigidity. (Statement I is false.) The relative rigidities of the two shear walls and the steel frame are not equal. (Statement III is false.)

The deflections of the two shear walls and the steel frame are equivalent to the deflection of the entire wall. (Statement IV is true.)

The answer is (B).

Customary U.S. Solution

The absolute rigidity of the entire wall is

$$R_{abs,entire\,wall} = R_{abs,masonry\,wall} + R_{abs,steel\,frame}$$
$$+ R_{abs,concrete\,wall}$$
$$= 15 \text{ kips/in}$$

The masonry shear wall, steel frame, and concrete shear wall resist the lateral load in proportion to their relative rigidities. The total lateral load is

$$V_{total} = V_{masonry\,wall} + V_{steel\,frame} + V_{concrete\,wall}$$
$$= 40 \text{ kips} + 15 \text{ kips} + 20 \text{ kips}$$
$$= 75 \text{ kips}$$

For the masonry shear wall,

$$V_{\text{masonry wall}} = \left(\frac{R_{\text{abs,masonry wall}}}{R_{\text{abs,entire wall}}} \right) V_{\text{total}}$$

$$R_{\text{abs,masonry wall}} = \frac{V_{\text{masonry wall}} R_{\text{abs,entire wall}}}{V_{\text{total}}}$$

$$= \frac{(40 \text{ kips}) \left(15 \frac{\text{kips}}{\text{in}} \right)}{75 \text{ kips}}$$

$$= 8 \text{ kips/in}$$

For the steel frame,

$$V_{\text{steel frame}} = \left(\frac{R_{\text{abs,steel frame}}}{R_{\text{abs,entire wall}}} \right) V_{\text{total}}$$

$$R_{\text{abs,steel frame}} = \frac{V_{\text{steel frame}} R_{\text{abs,entire wall}}}{V_{\text{total}}}$$

$$= \frac{(15 \text{ kips}) \left(15 \frac{\text{kips}}{\text{in}} \right)}{75 \text{ kips}}$$

$$= 3 \text{ kips/in}$$

For the concrete wall,

$$V_{\text{concrete wall}} = \left(\frac{R_{\text{abs,concrete wall}}}{R_{\text{abs,entire wall}}} \right) V_{\text{total}}$$

$$R_{\text{abs,concrete wall}} = \frac{V_{\text{concrete wall}} R_{\text{abs,entire wall}}}{V_{\text{total}}}$$

$$= \frac{(20 \text{ kips}) \left(15 \frac{\text{kips}}{\text{in}} \right)}{75 \text{ kips}}$$

$$= 4 \text{ kips/in}$$

The masonry wall has the largest absolute rigidity of 8 kips/in. (Statement II is true.)

The relative rigidities of the two shear walls and the steel frame are

$$R_{\text{rel},i} = \frac{R_{\text{abs},i}}{\sum_i R_{\text{abs},i}} = \frac{R_{\text{abs},i}}{R_{\text{abs,entire wall}}}$$

$$R_{\text{rel,masonry wall}} = \frac{R_{\text{abs,masonry wall}}}{R_{\text{abs,entire wall}}} = \frac{8 \dfrac{\text{kips}}{\text{in}}}{15 \dfrac{\text{kips}}{\text{in}}} = 0.533$$

$$R_{\text{rel,steel frame}} = \frac{R_{\text{abs,steel frame}}}{R_{\text{abs,entire wall}}} = \frac{3 \dfrac{\text{kips}}{\text{in}}}{15 \dfrac{\text{kips}}{\text{in}}} = 0.200$$

$$R_{\text{rel,concrete wall}} = \frac{R_{\text{abs,concrete wall}}}{R_{\text{abs,entire wall}}} = \frac{4 \dfrac{\text{kips}}{\text{in}}}{15 \dfrac{\text{kips}}{\text{in}}} = 0.267$$

$R_{\text{rel,steel frame}} < R_{\text{rel,concrete wall}} < R_{\text{rel,masonry wall}}$, so the steelframe does not have the largest relative rigidity. (Statement I is false.) The relative rigidities of the two shear walls and the steel frame are not equal. (Statement III is false.)

The deflections of the two shear walls and the steel frame are equivalent to the deflection of the entire wall. (Statement IV is true.)

The answer is (B).

87. Bridge columns with inadequate stiffness experience cracking of the members during earthquakes. The first step in the evaluation of deficient columns is to determine if and where plastic hinging will occur. Usually, plastic hinges are found in the end regions of columns or the footings. Steel jacketing (casings) is used to provide adequate confining pressure on the columns in the plastic hinge zones. There are different types of steel casings depending on the retrofit analysis.

Full-height column casing retrofit: A full-height column casing retrofit with cut-off at the pile transition will provide column confinement and will result in greatly increased column shear strength. The steel jackets or shells will provide confining pressure on the columns in the plastic hinge zones. When a poorly detailed column (i.e., minimal confinement steel and lap splices at the tops and the footings) is retrofitted with steel jackets or shells that provide adequate confining pressure, its shear strength will increase greatly. However, to effectively improve the column's strength, the small space between the column and the solid-steel shell should be grouted solid.

Partial-height column casing retrofit: Any location where a plastic hinge is assumed to form should be retrofitted. For example, for a fixed-end condition, the partial-height column casing retrofit at the top will provide column confinement and will permit a pin to form at the base. As a result, column moments will shift to the column tops where the column steel jacket (casing) will provide the needed ductility. This method will increase moment ductility and column shear strength.

Seismic anchor slab: The seismic anchor slab is used in a bridge structure retrofit to substantially stiffen the abutments. However, it is not an appropriate option for a bridge with inadequate columns, because the seismic anchor slab retrofit strategy draws larger seismic forces to the abutments. Seismic anchor slabs resist both longitudinal and transverse seismic displacements at each abutment. Each abutment should be evaluated for compression effects (e.g., bridge moving toward fill—anchor slab and abutment diaphragm activate large soil wedge) and tension effects (e.g., bridge moving away from fill—anchor slab is pulled across fill).

The answer is (B).

88. ASCE/SEI7 Sec. 11.2 defines moment-resisting frames. These frames resist forces in members and joints primarily by flexure, and rely on the frame to carry both vertical and lateral loads. The lateral loads are carried by flexure in the members and joints, and joints are theoretically completely rigid. Moment frames can be steel, concrete truss, or composite moment frames. They can also be categorized into ordinary moment frames (OMF), intermediate moment frames (IMF), and special moment frames (SMF). ASCE/SEI7 Table 12.2-1 notes the limits for the usage of these systems.

The answer is (C).

89. IBC Table 2306.3(1) gives a 170 lbf/ft (2480 N/m) allowable shear stress for 7/16 in (11 mm) structural I panels that are applied directly to framing and stapled with $1\frac{1}{2}$ 16 gage staples at 6 in (152 mm) o.c. at panel edges, and at 12 in (305 mm) o.c. along intermediate supports.

The answer is (C).

90. ASCE/SEI7 Sec. 20.3.1 defines site class F as

1. soils vulnerable to potential failure or collapse under seismic loading (e.g., liquefiable soils, quick and highly sensitive clays, and collapsible weakly cemented soils)

2. peats and/or highly organic clays ($H > 10$ ft (3 m)) of peat and/or highly organic clay where H is the thickness of soil

3. very high plasticity clays ($H > 25$ ft (7.6 m) with PI > 75)

4. very thick soft/medium stiff clays ($H > 120$ ft (37 m)) with $s_u < 1000$ lbf/ft^2 (50 kPa) in a soil profile that would otherwise be classified as Site Class D or E, with exceptions as outline in ASCE 7.

Where the total thickness of soft clay is greater than 10 ft (3 m) with $s_u < 500$ lbf/ft^2 (25 kPa), w_{mc} 40%, and PI > 20, the site class is defined as E, not F [ASCE/SEI7 Sec. 20.3.2].

The answer is (C).

91. ASCE/SEI7 Table 12.6-1 requires all irregular structures with horizontal irregularity type 1a to be analyzed using the modal response spectrum analysis or the seismic response history procedure.

The answer is (D).

92. In general, retrofit costs fall into two main categories: direct and indirect costs. The direct costs represent the sum of money a property owner spends to improve the building's deficiencies against earthquakes. An example of direct costs is a bill that a building contractor charges a property owner. The indirect costs arise as a result of a decision to proceed with the improvement of the building's deficiencies against earthquakes. Examples of indirect costs are construction permits, fees, and construction financing.

A building's structural system exerts the greatest direct influence on the amount of work required for seismic rehabilitation. Other factors can be location and the level of rehabilitation (retrofit) involved. Costs based on the building's structural system recognize the range of deficiencies inherent in a building's structural system.

For example, the typical seismic rehabilitation cost for frames with infill masonry is higher than the seismic rehabilitation cost for wood structures. The greater mass of the masonry building results in greater inertia forces that must be resisted. The perimeter of the masonry infill needs to be anchored to the frame to reduce the possibility of falling debris in the event of an earthquake. Furthermore, the building deflections need to be significantly decreased to reduce the loads unintentionally resisted by the infill material. The decrease in building deflection can be accomplished by increasing the building stiffness. All these seismic rehabilitation measures result in higher costs.

There is little or no correlation between the cost per square foot (square meter) and the height of the building or the building roof type. The building material influences the rehabilitation costs, but this is mostly accounted for by the structural system.

The answer is (A).

93. A post-and-pier foundation consists of posts that support the entire or portions of the building and that are, in turn, supported on isolated concrete piers (footings). This type of foundation system does not have continuous perimeter foundations or a substantial bracing system to resist earthquake forces. Although many houses on steep hills are built with this type of system, they are vulnerable to earthquakes unless specially engineered for the site.

For posts on piers, the top of every wood post should be reinforced at its connection to the beam with a steel strap on both sides of the post-and-beam connection. Steel T-straps or steel ledgers are preferred. In order to connect post and pier together where the post rests on the pier, a sheet-metal connector should be used on both sides of the post-and-pier connection.

The sheet-metal connectors should be pre-drilled. To reinforce the post-and-pier connections properly when using a sheet-metal connector, posts should be secured with bolts and piers should be connected with anchor bolts.

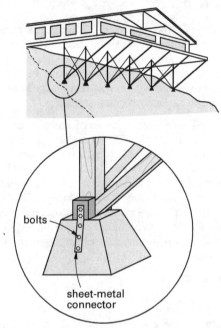

Additionally, it is important to brace wood posts with double X-diagonal braces around the perimeter of the structure and, at a minimum, every second line of interior posts, in each direction. The introduction of X-bracing provides a minimum amount of resistance to lateral forces. An even better solution is to add more posts and attach structural wood panels to create a shear wall bracing system.

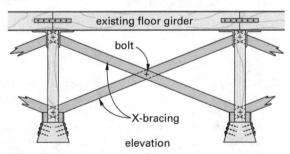

The answer is (D).

94. ASCE/SEI7 Sec. 12.12.5 describes deformation compatibility. It states that for structures assigned to seismic design categories D through F, every structural component not included in the seismic force-resisting system in the direction under consideration must be designed to be adequate for the gravity load effects and the seismic forces resulting from displacement to the design story drift.

The answer is (C).

95. ASCE/SEI7 Table 12.2-1 limits the use of special reinforced concrete shear walls to buildings no greater than 160 ft high when their seismic design categories are E. However, the building height may be increased to 240 ft if both of the given conditions are met [ASCE/SEI7 Sec. 12.2.5.4].

1. The structure does not have a type 1b horizontal structural irregularity.

2. The shear walls in any one plane resist no more than 60% of the total seismic forces in each direction, neglecting accidental torsional effects.

ASCE/SEI7 Table 12.2-1 does not prescribe any limits for buildings braced using a dual system, even though engineers routinely add special concrete moment frames that are capable of resisting at least 25% of the prescribed seismic forces.

The answer is (C).

96. For a flexible diaphragm, the lateral loads will be transferred to the supplemental elements in proportion to their tributary areas. The addition of supplemental elements by alternative I will primarily reduce the forces on the existing braced elements in the plane where the braced elements are added. For flexible diaphragms, this alternative essentially has no effect on the forces in braced elements located in other bays or on the diaphragms or the connections between the diaphragms and the vertical-resisting elements, but it will still reduce the seismic demand on the bracing elements. The addition of supplemental elements by alternative II will reduce the forces on all the elements, including existing braced elements, the diaphragm, the foundation, and connections. For flexible diaphragms, the force will be proportional to the tributary areas associated with the braced elements. Alternative II creates subdiaphragms with smaller tributary areas.

In reducing forces on overstressed elements or components, the rigidity, strength, and ductility compatibility of the existing vertical-resisting elements relative to the new vertical-resisting elements must be ensured. Both alternatives will reduce the seismic demand on the bracing elements, but alternative II will result in a much more resistant structure.

The answer is (C).

97. There is an immediate need for building safety evaluations and damage inspections after an earthquake strikes a community. Unsafe buildings should be identified, and people need to be kept from entering or using these buildings. Local building departments may seek additional qualified individuals for inspection when the extent of the earthquake disaster is great.

Building safety evaluation inspectors do not have to be structural engineers. Typically, they are qualified building inspectors. These individuals are familiar with building construction but do not necessarily have the experience of structural plan checkers or structural engineers. Civil and structural engineers and architects can also perform structural damage inspections.

This specificity is necessary to exclude other disciplines of engineering, such as nuclear, chemical, and mechanical, that are not experienced.

The answer is (D).

98. Foundations with piles or drilled piers that lack lateral force capacity to adequately transfer the seismic shears from the pile caps and the piles to the soil (ground) can be improved by

  removing the existing pile caps, driving additional piles, and providing new pile caps of larger size.

  reducing loads on the piles or piers. This can be done by providing supplemental vertical-resisting elements (i.e., shear walls or braced frames) and transferring forces to other foundation members with reserve capacity.

  adding tie beams to adjacent pile caps. This measure causes the loads on pile caps to be reduced and distributed.

IBC Chap. 18 contains additional information about foundations.

The answer is (D).

99. ASCE/SEI7 Sec. 13.3 provides the requirement for the seismic demands on nonstructural components. These demands are imposed on elements of structures and their attachments, permanent nonstructural components and their attachments, and the attachments for permanent equipment supported by a structure, which should be designed to resist the total design seismic forces, F_p, obtained from ASCE/SEI7 Eq. 13.3-1, Eq. 13.3-2, and Eq. 13.3-4. F_p should not be less than $0.3 S_{DS} I_p W_p$ or more than $1.6 S_{DS} I_p W_p$.

The answer is (C).

100. Requirements for the seismic design of mechanical and electrical components are given in ASCE/SEI7 Sec. 13.6. HVAC ductwork requirements are given in ASCE/SEI7 Sec. 13.6.6, which states that as long the component importance factor, I_p, is 1.0, seismic supports are not required if one of the given conditions is met.

  condition 1: HVAC ducts are suspended from hangers 12 in (305 mm) or less in length.

  condition 2: HVAC ducts have a cross-sectional area of less than 6 ft^2 (0.557 m^2).

From the problem, case I—HVAC ducts suspended from hangers 6 in (153 mm) long—satisfies condition 1. Case II—HVAC ducts having a cross-sectional area of 4 ft^2 (0.372 m^2)—satisfies condition 2.

Requirements for HVAC components are also discussed in ASCE/SEI7 Sec. 13.6.6. The code requires components, such as fans that weigh more than 75 lbf (334 N), to be braced.

The answer is (D).

101. For seismic resistance in a building, an adequate, complete, and sufficiently strong load path is a requirement. There should be a lateral-force resisting system that forms a direct load path consisting of elements within and between the vertical resisting elements, diaphragms, and foundations subsystems.

The lateral seismic inertia forces of an existing building are transferred from the floors and roofs through the vertical-resisting elements (e.g., shear walls, braced frames, and moment frames) to the foundations and into the ground. Gaps in the path cause a structure to fail under wind load and seismic forces. Examples of gaps are a discontinuous chord because of a notch in the diaphragm, a missing collector, or a connection that is not capable of transferring a diaphragm shear to a shear wall or frame.

The answer is (D).

102. *SI Solution*

ASCE/SEI7 Eq. 12.8-7 is used to solve for the approximate fundamental period, T_a. C_t and x are approximate period parameters determined from ASCE/SEI7 Table 12.8-2. h_n is the actual height of the building above the base on the nth level. For the structure given in the problem, and using ASCE/SEI7 Table 12.8-2,

$$T_a = C_t h_n^x = (0.0724)(15 \text{ m})^{0.8}$$
$$= 0.63 \text{ s} \quad (0.6 \text{ s})$$

The answer is (D).

Customary U.S. Solution

ASCE/SEI7 Eq. 12.8-7 is used to solve for the approximate fundamental period, T_a. C_t and x are approximate period parameters determined from ASCE/SEI7 Table 12.8-2. h_n is the actual height of the building above the base on the nth level. For the structure given in the problem, and using ASCE/SEI7 Table 12.8-2,

$$T_a = C_t h_n^x = (0.028)(50 \text{ ft})^{0.8}$$
$$= 0.64 \text{ sec} \quad (0.6 \text{ sec})$$

The answer is (D).

103. Both chord and collector elements are parts of the diaphragm of a structure. A chord operates as the tension strength for the diaphragm. A collector runs along the vertical plane of a wall or frame to collect diaphragm shear. Diaphragm materials (e.g., concrete slabs, plywood) are typically strong in compression, but weak in tension. Both chords and collectors are mostly for tensile loading. A single structural element (e.g., wood-frame top plate, reinforcing steel, steel floor beams) may operate as both members, depending on the direction of the lateral load. Generally, collectors accumulate force from the diaphragm along the walls and frames parallel to the direction of loading. The chord is the tensile element for diaphragm bending and is usually perpendicular to the direction of loading.

The answer is (B).

104. According to the exceptions listed in ASCE/SEI7 Sec. 12.7.2, floor live loads in public garages do not need to be included when calculating a structure's effective seismic weight.

The answer is (C).

105. An ordinary reinforced concrete shear wall paired with a special moment-resisting frame is a dual system. According to ASCE/SEI7 Table 12.2-1, for structures located in seismic design category D, ordinary shear wall dual systems are not permitted.

The answer is (B).

106. ASCE/SEI7 Table 12.6-1 identifies the permitted procedures by seismic design category. All methods may be used for buildings in seismic design category B. Seismic response history procedures may be used for any building. All methods may be used for buildings less than 160 ft and that have no structural irregularities. Therefore, all of the statements are true.

The answer is (D).

107. One of the requirements of ASCE/SEI7 Table 12.3-3 is that the effect of removing portions of the frame be considered. Item b of ASCE/SEI7 Sec. 12.3.4.2 relates to the location of lateral forces on the perimeter of the building.

While each floor's weight is needed to determine the magnitude of seismic loads, weight is not used to determine ρ.

The answer is (D).

108. *SI Solution*

Story drift, Δ, is the lateral deflection of one story relative to the story below. Use ASCE/SEI7 Eq. 12.8-15 to find the lateral deflections of stories 5 and 4. C_d is the deflection amplification factor; δ_{xe} is the elastic, design-level deflection; and I_e is the importance factor.

$$\delta_x = \frac{C_d \delta_{xe}}{I_e}$$

From ASCE/SEI7 Table 12.2-1, for a special moment-resisting frame, C_d, is 5.5.

The lateral deflection for level 5 is

$$\delta_5 = \frac{(5.5)(71 \text{ mm})}{1.0} = 390 \text{ mm}$$

The lateral deflection for level 4 is

$$\delta_4 = \frac{(5.5)(51 \text{ mm})}{1.0} = 279 \text{ mm}$$

The design story drift is the difference in deflections.

$$\Delta_{5/4} = \delta_5 - \delta_4 = 390 \text{ mm} - 279 \text{ mm}$$
$$= 112 \text{ mm} \quad (110 \text{ mm})$$

The answer is (C).

Customary U.S. Solution

Story drift, Δ, is the lateral deflection of one story relative to the story below. Use ASCE/SEI7 Eq. 12.8-15 to find the lateral deflections of stories 5 and 4. C_d is the deflection amplification factor; δ_{xe} is the elastic, design-level deflection; and I_e is the importance factor.

$$\delta_x = \frac{C_d \delta_{xe}}{I_e}$$

From ASCE/SEI7 Table 12.2-1, for a special moment-resisting frame, C_d, is 5.5.

The lateral deflection for level 5 is

$$\delta_5 = \frac{(5.5)(2.8 \text{ in})}{1.0} = 15.4 \text{ in}$$

The lateral deflection for level 4 is

$$\delta_4 = \frac{(5.5)(2.0 \text{ in})}{1.0} = 11.0 \text{ in}$$

The design story drift is the difference in deflections.

$$\Delta_{5/4} = \delta_5 - \delta_4 = 15.4 \text{ in} - 11.0 \text{ in}$$
$$= 4.4 \text{ in}$$

The answer is (C).

109. IBC Table 1705.3 states that placement of reinforcing steel requires periodic, not continuous inspection.

The answer is (C).

110. Per ASCE/SEI7 Sec. 11.6, the more severe seismic design category should be used. Based on ASCE/SEI7 Table 11.6-1 and an S_{DS} value of 0.40, the seismic design category is C. However, based on ASCE/SEI7 Table 11.6-2 and an S_{D1} value of 0.10, the seismic design category is B. The more severe seismic design category is C.

The answer is (C).